Biolegalities

Series Editors

This interdisciplinary series on *Biolegalities* engages with contemporary challenges and implications of new biotechnologies and biological knowledges in the field of law. Our series aims to open up a broader understanding of biolegality that includes a range of biotechnologies and biological knowledge, expanding into areas of immigration law, trade law, labor law, environmental law, patent law, family law, health law, human rights law, and international law. While the growing scholarship on biopolitics has studied the ways in which such practices are entangled with certain modes of governance and neoliberal economies, their translations, deployments, and reconfigurations in the realm of law or legal practice has been relatively understudied. The main objective of this book series is to provide a venue for the study of the complex and often contested ways in which biotechnologies or biological knowledges are reworked by, with, and against legal knowledge.

Sonja van Wichelen · Marc de Leeuw

Biolegality

A Critical Introduction

palgrave
macmillan

Sonja van Wichelen
School of Social and Political Sciences
University of Sydney
Sydney, NSW, Australia

Marc de Leeuw
School of Law, Society and Criminology
University of New South Wales
Sydney, NSW, Australia

ISSN 3005-0227 ISSN 3005-0235 (electronic)
Biolegalities
ISBN 978-981-99-8751-1 ISBN 978-981-99-8749-8 (eBook)
https://doi.org/10.1007/978-981-99-8749-8

Cover image: The cover image represents cells from Henrietta Lacks, an African American woman who passed away in 1951 from cervical cancer. The HeLa cells encompassed the first human cell line able to reproduce indefinitely and became used in research around the world. After it became known that the cells were collected at the John Hopkins Hospital in Baltimore without consent, debate erupted about ownership of human biological material and the ethical conduct of biomedical research as discussed in Chapter 3 of this book. Title: Scanning electron micrograph of just-divided HeLa cells. Zeiss Merlin HR-SEM. Source: National Institutes of Health (NIH) Date: 17 August 2011, 02:40:59

This Palgrave Macmillan imprint is published by the registered company Springer Nature Singapore Pte Ltd.
The registered company address is: 152 Beach Road, #21-01/04 Gateway East, Singapore 189721, Singapore

Paper in this product is recyclable.

For Luc and Nils

Prologue

One morning in the fall of 2017, during our sabbatical in Berkeley, our 8-year-old son asked what we were writing about. We answered that we were writing a book about law and science. "But law and science aren't the same things" he replied. "How so?" we asked. "Law is about the rules, and science is about experimenting." "Well," we said, "our book is about the rules of experimenting … and experimenting with the rules."

In a crude version, our response to him sums up well what we are trying to do in this book. We are bringing together two apparently disparate fields, namely law and biology, to meaningfully engage with some of the most pressing issues in society today: the confrontation between new bioscientific and biotechnological advances, and legality in contemporary societies. These technoscientific advances are rapidly changing society, and law—as one of our defining social institutions—is a key site for the social and cultural translation and mediation of bioscience's impacts. But law is not usually where we analyze the relationship between science and society. Typical contemporary scholars turn to bioethics when investigating these social changes. Since its conception in the US in the 1970s, bioethics developed into a significant institution in the Anglo-American world (see 211 in Poland 2005). Nowadays, bioethical committees and working groups are routine, if not required, in any endeavor involving bioscientific experimentation or application. These efforts have made important contributions to jurisprudence and legal practice, leading to substantive changes in the interpretation of common law.

Despite its importance, our call to study biolegality moves us beyond bioethics and its use of expert knowledge to deliberate over and prescribe rules for policy and legal process. Rather than focus on rulemaking by experts, in this book we concentrate on the changing understanding of foundational legal principles precipitated by bioscientific advances and applied technologies. We look, instead, at the *enactment* of ethics through practical legal knowledge, describing the role the practice and purpose of ethics performs in the making of biolegality (Van Wichelen 2019). These are not ideas or reconfigurations of ethical norms brought about by moral philosophers, ethicists, or medical experts, but by a contingent encounter between institutional practices, (trans)national imaginaries, and communal (be)longings. While the book concentrates on legal concepts—property, personhood, parenthood, (collective) identity, and so on—we approach these as social and political objects investigating how, and to what extent, biotechnology rearranges the original premises of law.

One could argue that these investigations fall within the category of biolaw, but even though we engage with legal theory and contemporary jurisprudence, our investigations are instead informed by the anthropology and philosophy of law, focusing more broadly on social ordering, which takes place inside and outside the confines of biolaw. According to some legal scholars, biolaw—which emerged from France in the 1990s—presupposes bioethics and the two are therefore inseparable (Kemp 1998). Dealing with developments in medicine and biotechnology, the field is evolving rapidly and represents a central node in biotechnology's application in society at large. While the book engages with some of the core issues in biolaw, we also draw on scholarship from outside of law to show how scientific and technological knowledge, experience, and legitimacy are performatively constituted through practice rather than through deliberation. Thus our conception of "biolegality" is the product of a myriad of disciplines, including sociology, anthropology, history, political science, philosophy, cultural studies, and science and technology studies. This diverse scholarship describes legalities, legal consciousness, and knowledge practices as performed by lay individuals, communities, cultures, or nation-states, rather than bioethicists, medical experts, lawyers, and judges per se.

The aim of this book, then, is, despite their significance, to step back from bioethics and biolaw and to look at the work of biolegality as the larger social and political field these specialized discourses contribute to but neither transcend nor control. By "legality" we do not refer to its literal meaning that denotes lawfulness or an observance to law. We refer, instead, to the experience and meaning people attach to law as they engage, avoid, or resist it (see p. 335 in Silbey 2005). As we demonstrate, the complexity of opposing narratives and paradoxical situations regarding law and the biosciences strengthens rather than weakens the law as a structure of social action. Our aim is to bring together scholarship in this new field in order to describe how the law surrounding the biosciences at once affects the activities of people and communities and is itself affected by the way people and communities incorporate, oppose, or negotiate new developments in bioscience and biotechnology. We see these narratives as forming new legalities around the idea of nature and biology.

But legality is also about legitimacy. For instance, the formal regulations regarding freezing and banking eggs and semen for research or reproductive purposes (described in Chapter 4) legitimize—i.e. make both lawful and socially acceptable—the temporary storage of biological substance to advance knowledge or create life (see p. 472 in Waldby 2015). Working through the ways people, communities, institutions, markets, and states respond to these new legal particularities, new legitimacies emerge to inform new kinds of politics. In this example, freezing human reproductive material becomes a form of potentiality imbued with hope and fear, but also rights and duties. These rights and duties are entangled with forms of reproductive citizenship (Carroll and Kroløkke 2018) that embrace the values of nuclear family life, romantic couplehood, and genetic progenies nurtured by the state. Biotechnological legitimacy is thus formed by how institutions adjudicate and sanction the politics of life.

Such politics of life bring our book close to the project of biopolitics, the governance of life processes and populations through the power of authoritative knowledge practices (Lemke 2011, Mills 2017, Rose 2006). Indeed, biopolitical strategies and their inherent relationship to legitimacy are central to our investigation. Nonetheless, our book is also a programmatic intervention in the scholarship on biopolitics, which has often overlooked the explicit realm of law and legality, prioritizing governance and governmentality rather than legitimacy and legality (Fassin 2009, Pottage 1998). Focusing explicitly on law, we demonstrate not only that the law is a fascinating terrain on which to explore biopolitics, but that the study of biolegality introduces a greater complexity to biopolitics. This complexity has much to do with legal form and the rules of engaging the law. While the biosciences and their attendant technologies often reinforce or are appropriated by existing forms and rules, our main contention is that they also disturb, deconstruct, and transform legal form and rule. This creates new conditions for thinking about law, conditions we hope to illuminate in the service of conceptualizing biolegality for future use.

Sonja van Wichelen
Marc de Leeuw

References

Carroll, K. and Kroløkke, C. 2018. Freezing for Love: Enacting 'responsible' Reproductive Citizenship Through Egg Freezing. *Culture, Health & Sexuality*, 20(9), pp. 992–1005.

Fassin, D. 2009. Another Politics of Life Is Possible. *Theory, Culture & Society*, 26(5), pp. 44–60.

Kemp, P. 1998. The Bioethical Turn. In Rendtorff, Jacob and Kemp, Peter (eds.), *From Ethics to Biolaw*, pp. 9–20. Copenhagen: Centre for Ethics and Law, University of Copenhagen.

Lemke, T. 2011. Biopolitics. In *Biopolitics*. New York University Press.

Mills, C. 2017. *Biopolitics*. London: Routledge.
Poland, S.C. 2005. Bioethics, Biolaw, and Western Legal Heritage. *Kennedy Institute of Ethics Journal*, 15(2), pp. 211–218.
Pottage, A. 1998. The Inscription of Life in Law: Genes, Patents, and Biopolitics. *Modern Law Review*, 61, p. 740.
Pottage, A. and Sherman, B. 2010. *Figures of Invention: A History of Modern Patent Law*. Oxford University Press.
Rose, N. 2006. *The Politics of Life Itself: Biomedicine, Power, and Subjectivity in the Twenty-First Century*. Princeton: Princeton University Press.
Silbey, S.S. 2005. Everyday Life and the Constitution of Legality. In Jacobs, M. and Hanrahan, N. (eds.), *The Blackwell Companion to the Sociology of Culture*, pp. 332–345. Oxford: Blackwell.
Van Wichelen, S. 2019. Moving Children Through Private International Law: Institutions and the Enactment of Ethics. *Law & Society Review*, 53(3), pp. 671–705.
Waldby, C. 2015. 'Banking Time': Egg Freezing and the Negotiation of Future Fertility. *Culture, Health & Sexuality*, 17(4), pp. 470–482.

Acknowledgments

This project began during one sabbatical and was completed during another. Institutional support—both financially and intellectually—was essential to this endeavor. We would like to thank the University of Sydney and the University of New South Wales for providing the means for our sabbatical leaves. In particular, our last sabbatical—during pandemic times—was only possible because of high-level institutional leverage. In 2017 we were based with the Center for the Study of Law and Society (CSLS) at the University of California, Berkeley, and we thank Jonathan Simon and Catherine Albiston for hosting us. Our first ideas were presented to this group, and we are thankful for all the questions and comments we received from the CSLS community. Our time in the Berkeley Hills was marked by our proximity to the places where biotechnology was in the making and where legal cases formed important precedents. It was the hightide of the CRISPR-Cas9 technique and we were fortunate to attend public talks and discussions led by one of its inventors, Jennifer Doudna. It was an exciting time and place to be thinking about the ethics and legalities of emerging biotechnologies.

It took another sabbatical in 2020–2021 to complete this project. This was amid the global COVID-19 pandemic and while Australia was locked out from the rest of the world—and therefore responding well to the threat of contagion—we managed to travel from Sydney to start our Fellowships with the Institute for Advanced Study (IAS) in Princeton, New Jersey. We would like to thank Didier Fassin and Alondra Nelson for inviting us to join their School of Social Science. Even during a pandemic, they managed to sustain intellectual debate and it was such an ironic coincidence that the 2020–2021 seminar focused on "Science and the State," a topic well-suited to pandemic times. Our thanks go out to Alondra Nelson and Charis Thompson for directing the seminar, whose discussions informed much of the thinking in this book. We also would like to thank all the other scholars of the 2020–2021 cohort, especially Joshua Barkan, Joy Rohde, Thomas Fossen, Magdalena Malecka, Jacob Foster, Leslie Paik, Nusrat Chowdhury, Banu Bargu, Lawrie Balfour, Christo Sims, Diana Graizbord, and Waqar Zaidi for their collegiality and friendship. And a special thanks to Miles Jackson from the School of Historical Studies for his continued support of our biolegalities project.

This project received generous funding and we would like to acknowledge the Institute for Advanced Study—especially the Institute's Founder Circle Members Deborah Lunder and Alan Ezekowitz. Sonja would also like to acknowledge the Sydney Research Accelerator (SOAR) prize scheme at the University of Sydney, which generously allowed her time away from teaching to write a significant part of the manuscript, and the Sydney Social Sciences and Humanities Advanced Research Centre (SSSHARC) for funding the Biolegality Pup-Up Research Lab, which brought scholars together from across the world to discuss biolegality. Marc would like to thank the UNSW Faculty of Law & Justice for supporting the UNSW Initiative for Biolegalities for several symposia ("The 'Biological Turn' in Law: A Critical Appraisal"; "The Seeds of Law: Monsanto, Patents and the Public Good"; and "Natural and Unnatural Threats: Pandemics & the Global Governance of Health and Biosecurity").

We would like to thank our editors Joshua Pitt and Marion Duval for their relentless support, enthusiasm, and patience in bringing this manuscript to production. Palgrave Macmillan's advocacy of Science and Technology Studies is encouraging for the discipline, and their support for bringing the Biolegalities Book Series into existence is a sign of their editorial willingness to take up projects at the margins of conventional disciplines. Ben Stork—our personal editor—meticulously read the whole manuscript and we thank him deeply not only for his highly professional editorial assistance, but also for his very constructive and helpful comments, which have significantly improved the quality and clarity of our manuscript. Thanks to Gopalakrishna from Springer Nature for her immense patience in the final production stage of the manuscript.

Finally, our sons Luc and Nils have endured two transcontinental sabbatical moves and too many talks around this never-ending book. We are proud of their dispositions to tag along and to trust us that—even in times that we were unsure ourselves—things will be okay. This book is dedicated to them.

Praise for *Biolegality*

"*Biolegality* offers a fascinating account of how advances in biotechnology shape our legal institutions and practices while simultaneously creating new ways for communities and societies to relate to biomedical knowledge, including genes, brains, bodies, and babies.

Drawing upon an impressive range of disciplines, they convincingly demonstrate how biotechnological knowledge transforms legal knowledge and entities, reconstructs the biomedical disciplines and nature itself, and shifts the social order. This is a must-read book."

—Myles W. Jackson, *Albers-Schönberg Professor in the History of Science, Institute for Advanced Study, Princeton, USA*

"*Biolegality. A Critical Introduction* offers an excellent account on the manifold constellations of law and biology in contemporary societies. Drawing on science and technology studies, political theory and legal anthropology, the book attends to the co-production of legal knowledge and biotechnological practices and convincingly shows how they challenge our understanding of the human and reconfigure concepts of property, personhood, kinship, community and identity."

—Thomas Lemke, *Professor of Sociology with Focus on Biotechnologies, Nature and Society, Goethe University, Frankfurt, Germany*

"Moving well past existing research that connects the separate regimes of law and the biosciences, this book illuminates their coproduction in the generative field of *biolegalities*. Van Wichelen and de Leeuw offer a significant theoretical advance in defining this compelling new paradigm. Drawing on many exemplars from the latest practical developments, they offer cutting-edge insight into the ecology of tensions and transitions in politics, life, biotechnology and law that continue to unsettle and shape the legalities surrounding human conditions of existence."

—Margaret Davies, *Matthew Flinders Distinguished Professor, Flinders University, Adelaide, Australia*

Contents

1

Introduction: Biolegality as Critical Intervention

Welcome to a CRISP(e)R World

In 2012, microbiologists Emmanuelle Charpentier and Jennifer Doudna made a discovery that would usher into existence the world of gene editing.[1] They had discovered how an enzyme (Cas9) in combination with CRISPR (Clustered Regularly Interspaced Short Palindromic Repeats)—a bacterial immune system—could cleave to specific parts of DNA, demonstrating this system can be adapted to make targeted cuts in a genome, modifying a DNA sequence. Although still in its early stages, the technique is heralded for its possible future uses in various domains, including all kinds of genomics, disease models, and anti-microbials, as well as agricultural, food, and biotechnological applications (Barrangou and Doudna 2016). As a genome editing tool, CRISPR-Cas9 transformed the scientific community because it is more accurate, efficient, and cheaper than other genome editing techniques.

As labs around the world experiment with gene editing, they are, among other things, editing whole organisms to study disease resistance, changing wheat to more effectively resist pests, altering chromosomes and mutations in mice to tackle cancer and hereditary disease, and editing the DNA of pigs to grow organs for transplantation. But the most

S. van Wichelen and M. de Leeuw, *Biolegality*, Biolegalities,
https://doi.org/10.1007/978-981-99-8749-8_7

divisive application is that of human germline gene editing; the heritable alteration of germ cells (those that grow into eggs and sperm). As the scientific community advises strict caution, laws and regulations across the world prohibit the use of germline editing for reproduction. Yet, in 2018, the Chinese scientist He Jiankui used the technique to edit human embryos, with the intent, according to He, to eradicate the potential HIV infection for these babies and their offspring.[2] With this event, the world witnessed the first "CRISPR'd babies" (Greely 2019), a move seen by many as tremendously controversial and having far-reaching consequences.[3]

Like with all other new biotechnologies (one only needs to think back to the first IVF baby) gene editing is accompanied by thrill as well as fear. The thrill consists of the promise that the science will, for instance, eradicate disease more effectively, or provide better and more nutritious crops to feed the growing world population.[4] But there is also fear that the technology will bring calamity and disaster to humanity, society, and the environment. The case of He Jiankui prompted wide public discussion whether gene editing in humans will foster new eugenic practices through the creation of "designer babies." Changes made to the human germline are also heritable, affecting future generations, and raising questions about how gene editing will shape humanity's future. This technique asks us to consider, how to deliberate and manage the idea of eliminating unwanted traits or enhancing desirable ones? How gene editing, by tinkering and interfering with nature, might enhance biological warfare or extinguish species and ecologies (Andorno et al. 2020; Braverman 2017; Doudna and Steinberg 2017)?

In responding to these fears, society calls upon bioethics, and the legal field calls upon biolaw, to deliberate and bring clarity to the matters at hand. Established in accordance with various national priorities and from an existing pool of experts, bioethical committees provide recommendations that inform institutional guidelines and regulation. The response to He exemplified the centrality of moral and ethical deliberation by experts; rules are to be formed from these deliberations, demarcating boundaries of how far scientists should go in editing human genetic material. The field of biolaw—developed primarily in continental Europe since the 1990s—complements bioethics by incorporating bioethical principles within the language and institutions of law. Bioethics—as the philosopher Peter Kemp argues—cannot be disentangled from biolaw, for bioethical principles inform how biolaw should be shaped: "bioethics is the ethics of the body, and the ethics of the body is the foundation for biolaw" (see Kemp quoted on p. 215 in Poland 2005).

Our approach to biolegality is not an attempt to replace bioethics or biolaw. Nor is it the constitution of a particular regime of biolaw and bioethics that ultimately shapes how they relate to one another, i.e. the ways that certain bioethical norms are codified in biolaw and the ways applications of biolaw (outcomes) refine bioethical norms through existing practices. Rather, biolegality exists alongside bioethics and biolaw as a useful conceptual device to address how biotechnological advances and practices affect our existing legal institutions, forms, and knowledge, while at the same time, constructing for people, communities, and societies, new or altered forms of relating to biological and biotechnological knowledge. This means that biolegality refers as much to the *outcome* of bioethics and biolaw as to that which *informs* it in the first place. While conventional bioethics and biolaw is vital for the stimulation of public debate and to changing existing, or forming new, regulations and law around biotechnologies, the study of biolegality illuminates two core entanglements of biotechnology and law that the twin concepts of bioethics and biolaw fail to account for: (1) the political dimensions regarding legal governance and (2) the knowledge practices informing legal form.

First, using the case of gene editing we started this introduction with, we can see the ways mainstream bioethics—especially the Anglo-American strand—fails to explain the political dimensions of emerging biotechnologies, in effect depoliticizing the institutional interests underpinning contemporary problems in biology and law, thereby naturalizing the biopolitical (see p. 13 in Swiffen 2010, see also Obasogie and Darnovsky 2018). By naturalizing the biopolitical, we mean the ways in which modern law since the nineteenth century, and Western (US) bioethics since the 1970s, have framed the organization of political life as conditioned upon certain bodies and populations and their capability to be (socially) healthy, thereby extending the "make live, let die" ethos—onto apparatuses of (neo)liberal governmentality. Indeed, such depoliticization can ironically be understood as a politics of bioethics (Petersen 2011), or more aptly as the bioethical "politics of anti-politics" (Brown 2002). From the very beginning, gene editing evoked a deep worry among the general public about its implications. However, as Sparrow and Mills explain, Anglo-American bioethics "struggles to articulate this unease or indeed to discern much wrong at all with using CRISPR-Cas9 to genetically modify human beings in vitro" (see p. 2 in Sparrow and Mills 2021). There seems to be a gap, then, between what biotechnologies evoke in people's imagination, and the way in which bioethics can readily engage these thoughts.

The expansive study of biopolitics has successfully addressed this gap.[5] By studying the political legitimacies behind existing and desired regulatory frameworks and governance systems—themselves often conditioned by power differentials and the interests of states and markets—we can better understand what the stakes are in debates around bioscientific developments like gene editing and how regulatory policies address these stakes. The relationship between law and bio(techno)logy, however, assumes its own dynamic that must be differentiated from the operation of biopolitics. Bioethics and biolaw play a particular role in the constitution of the biopolitical and biolegality is a way of explaining this role, first, by bringing back the political into the ethico-legal appraisals of biotechnology, and second by focusing more explicitly on the material artefacts produced by regulatory institutions in their interactions with bioscientific knowledge (Lezaun 2006). While biopolitics critically examines the normative and ideological dimensions informing policy and governance (including legal instantiations), biolegality offers a lens through which to understand the constitution of biolegal objects more specifically, namely those molecularized or bioengineered normative forms that have the capacity to reorder social life and their institutions (Jasanoff 1997, 2001; Lezaun 2006; Lynch and McNally 2009; Lynch et al. 2010; Pottage 2007).

Second, recent developments in the life sciences focus on the *making* in addition to the *knowing* of nature and biology (Rabinow 1992,

2010), a shift epitomized by the example of gene editing but also of synthetic biology more generally. During the 1990s the focus of the Human Genome Project was to map the sequences of human DNA and the genes of the human genome. The emphasis was on what kind of information this mapping provides and how this information might assist in the development of (agricultural or biomedical) technologies and therapeutics. In contrast, the turn to slicing and editing assumes control over heritability, thereby providing scientists the capacity to change specific traits of any organism. The anthropologist Paul Rabinow observed this turn with the advent of synthetic biology—the bioscientific branch involved in the engineering of organisms—and marked it as the *enculturalization* of nature (1992, 1998, 2010).[6] Of course, nature in itself is a social and historical construction, but the way in which we have related to nature, particularly in modernity, is as if it is ahistorical, universal, and given (Latour 1999). Rabinow argues that bioscientific developments such as synthetic biology (and we can now include gene editing) are changing the premises of this construction. As Rabinow shows, objections against the "unnaturalness" of GMOs (or artificialness of gene editing) pose a challenge and opportunity, rather than a threat to, or the end of biotechnology; it is not inconceivable that the genetically altered human becomes the desired biology, just as genetically modified tomatoes have become the desired natural tomato.

The shift from knowing to engineering affects the knowledge practices that inform legal form. Bioethics and biolaw struggle to address these changes in knowledge practices. As we describe more fully below, this is because their methodological frameworks—attached to either prescriptive normative and philosophical principles or formal and doctrinal law—cannot account for the legitimating systems or ideas that accompany the changing knowledge practices taking place across institutions, people, communities, and societies in the context of bioengineering. As a conceptual tool working alongside bioethics and biolaw, biolegality can help examine how bio(techno)logical advancement in the biopolitical age engages, troubles, and rearranges legal form and reorders the construction of biology or nature, and what this reordering generates in the realm of knowledge and sociality.

This approach goes against the common assumption in media and public discourse that law *lags* behind science and technology, and that "catching-up" is what law needs to do to render its instruments compatible with the latest developments—both in terms of efficiency as well as ethics. In our example of gene editing, the "law lag" principle assumes that the practical and ethical questions around gene editing will be settled once law—with the help of bioethics—catches up to regulate the new technology. But bioscientific developments do not occur in a legal vacuum. On the contrary, the production of the gene editing technique itself, as well as the ensuing debates regarding what is permissible, are both conditioned by prevailing legal norms, those concerned for instance with research ethics, intellectual property, and the integrity of human dignity and biological diversity. Courts play a central role in integrating technological developments into society (Jasanoff 1997: 180) and push actors to revisit existing regulations. They evoke public discussion on the question how and to what extent biotechnological advances are to be integrated in our biological, social, and institutional lives. The Science and Technology Studies (STS) scholar Sheila Jasanoff argues in this respect, that instead of the "law lag" principle, law and science more aptly follow the process of "co-production" (2001). Here, epistemic, and normative understandings of life are reordered by law and science in a recursive fashion: one does not follow the other, but are mutually constituted. While the bioethical premise informing a new technique such as CRIPR-Cas9 necessarily assumes the well-rehearsed thesis that "law lags behind science," the *biolegal* premise points to the coproduction of biology and law. Here, law works from an internal dynamic, though not cut off from other legitimating systems.

To ilustrate, following the outrage around He Jiankui's experiment, the emerging literature on human germline gene editing describes the immense difficulties for institutions grappling with how to set up new rules governing experimentation. The Dutch legal scholar Britta Van Beers, however, rightly emphasizes that within the Euro-American context, legal frameworks around heritable human genome editing already existed since the late 1990s. Based on human rights discourse, these regulations either outright banned human genome editing, as in European countries, Australia, Canada, and Brazil, or allowed some form of the practice, though chiefly for research purposes, as in the US, the UK and China (Van Beers 2020: 8). The ethical or non-ethical practice of gene editing relies heavily on these already existing regulatory frameworks (governance of biomaterials, regulations around GMOs and LMOs, regulations around reproductive technology and cloning, etc.). Again, the idea that "law lags" scientific or technological accomplishments is inadequate; instead, these accomplishments develop through, against, or because of existing laws. The demand for regulation has, apparently, not been evoked by a lack of regulation, but by a more complex system of biolegal imaginaries and practices, emerging from scientific communities or publics and their desires for, or fears, of engineerable societies.

Gene editing troubles the modern and genetic stability of heredity by introducing a more extreme, more efficient, and more affordable way of altering life than the "genetic modification" of the recent past. The ensuing debate in Euro-American contexts, then, is how to regulate the science and industry that are altering crops and other biological material (including embryos and fetuses) through genome editing. In the case of human germline gene editing, regulatory calls are moving toward regimes of self-regulation by scientific communities. Critical observers see such a move as problematic for several reasons, though most clearly because such regimes exclude social and public participation (Greely 2019; Jasanoff; Hurlbut and Saha 2015; Sparrow and Mills 2021; Van Beers 2020). Epistemologically, gene editing disturbs the stability of knowledge premised on the principle that we cannot change the genetic makeup of biological life. Ontologically, it is remaking life itself. The legal premises through which gene editing occurs and the legalities that ensue through its practices will generate new knowledge as well as new subjectivities and socialities. These social realities need the attention of publics and citizens and bring back the political, thereby refusing the anti-politics of bioethics.

Of course gene editing is just one of the developments in contemporary bioscience that are complicating issues emerging at the intersection of law and biology, just one item in this long but still partial list:

gene drives, genetic privacy, genetically and living modified organisms (GMOs and LMOs), biobanks, transgenic animals, nanotechnology, neuro-interventions, hormonal therapies, cryopreservation, preimplantation genetic diagnosis (PGD), and mitochondrial replacement techniques. Moreover, these developments will eventually impact a wide-ranging area of law including—though not limited to—administrative, constitutional, contract, corporate, criminal, environmental, family, health, intellectual property, international law, tort, and human rights law. Our aim is not to give an exhaustive account of all these changes. Rather, this book singles out particular "slices of life" or "lively things" (see p. 16 in Winickoff 2015)—of *genes, brains, babies, bodies,* (and *viruses* in the epilogue)—to describe how biotechnology's reworking of law effects anthropological understandings of exchange, self, kinship, and community.

In the rest of this introduction, we advance our approach to biolegality and argue for its significance in understanding the entanglement of biology and law in contemporary society. We start with a discussion around how biolegality moves away from both conventional bioethical frameworks and scholarship in biolaw based in a doctrinal understanding of legal processes. Biolegality finds its affinity with biopolitical frameworks. However, as we describe the original context in which the term was born, namely in the field of forensics and criminal law, we place the term in a broader historical and empirical context of biology and law. We demonstrate how studies in biolegality can assist us beyond formal legal and ethical inquiries and contribute more forcefully to understanding the antagonistic—though intimate—relationship between the political dimensions of law and the knowledge practices informing legal and cultural forms. In the penultimate section we reflect on the question of temporality and assess whether these shifts in the relationship between law and biology can be designated as "new." Rather than signaling an epochal shift, we argue that the changes resemble "tipping points" (Tsing 2005) that change how law and life understand their engagement with the social; with the "wholes and the parts" of society (Strathern 1992). We end with a summary of the chapters.

Making Legality Legible

The term biolegality was first coined by the Science and Technology Studies scholars Michael Lynch and Ruth McNally (2009) to describe the coproduction of biotechnology and legislation within the context of forensic science, a field redefined in the past few decades by genetic technologies. As Lynch and McNally show, law and law enforcement play a unique role in communicating the power of genetics in criminal law. Their study demonstrates how the use of genetic evidence rearranges values, standards, and requirements within the criminal justice system (see also Lynch 2003 and Jasanoff 1997). The increasing use of DNA profiling as evidence in criminal courts produces certain kinds of suspects, who become the focus of scrutiny based solely on the status of their biological material or information. The coproduction of law and science in the operation of biolegality, they contend, constructs DNA as a "machinery of truth," whereas careful analysis of the contexts in which this truth is produced illuminates the fallacies of the use of genetic evidence. This exceptional reliance on DNA as a "truth machine," Lynch and McNally argue, indicates that biolegality is not only a "historical relationship between biological innovation and enabling legislation, [rather,] it is an epistemic relation in which biological 'truth' justifies exceptional legal procedures" (see p. 296 in 2009). Our book stays true to—and draws directly from—this understanding of biolegality but places it in a broader history of legal theory and expands it to other empirical realms of science and law, beyond the use of genetic evidence in forensic science and criminal law.[7]

It is not surprising, that many legal and social science scholars in the social study of science and technology rely on Foucault's framework for biopolitics and governmentality. Foucault's conceptualization of biopolitics draws explicitly on how knowledge from the life sciences—particularly biological knowledge—allowed life to be calculable (through truth regimes) in such a way that it became instrumental in the regulation of populations, such as the organization of the mentally ill, the HIV infected, the obese, etc., and in forms of subjectivation, for instance how—prompted by scientific knowledges—people change, alter, modify, nudge, perfect, emulate their selves and their bodies (see Lemke 2011 and Hacking 2006). Like Lynch and McNally, our approach recognizes the centrality of a Foucauldian framework of biopower in understanding the operations of biolegality. Many studies concerning the uses of biotechnologies in the governance of people deploy similar biopolitical perspectives. Immigration is a point in case, where governance and surveillance systems increasingly use biotechnology to control the flow of people coming into states. DNA kinship testing for the purpose of immigration similarly relies on the idea that DNA evidence promises "true" knowledge of the body and heredity (see p. 30 in Heinemann and Lemke 2014).[8] Likewise, technologies of border security have been further refined with biometrics and (epi)genomic technologies (Ihar 2020; Pugliese 2012; M'charek et al. 2014). The use of such technologies to advance border security shows the intricate and political entanglement of biolegality with what has been termed "biocitizenship." Here, calculable, quantifiable, and ultimately depoliticized biomarkers become useful entities for the governance and administrative processing of people and populations. The biopolitical framework allows these scholars to focus on the epistemic shifts that rearrange the value and scrutiny of life itself, while enabling, simultaneously, a critical appraisal of the juridico-political structures of modern and liberal society.

As an empirical or historical dynamic, biolegality can only be understood within the larger significance of a biopolitical regime that replaced more repressive forms of sovereignty from the twentieth century onwards. As Lynch and McNally underscore in their original contribution to the concept, the scholarship on biopolitics provides a deeper understanding as to how biolegal conditions have come to occupy the place they do today. Foucault formulated the term to demonstrate a shift in sovereign power from the sovereign that "makes die and lets live" (e.g. on the executioners scaffold where the subject's life or death is directly determined by sovereign power) and a disciplinary power organized under the rubric of "to make live and to let die," which constitutes a regime of biopolitics that takes hold of "life itself" as the object of governance. The state's primary occupation became to regulate its populations in order to "ensure, sustain, and multiply life, to put this life in order" (see 138 in 1976: 138). Here, biopolitics identifies and analyzes the strategies through which life is managed, scrutinizing the institutions of knowledge and power by unfolding how they form new subjects and socialities. One could then say that biolegality is a technique or apparatus of biopolitics.

Yet, to describe biolegality as merely another form of biopolitics would reduce the complexity of its operation. Law acts as an autopoietic system, meaning that it depends on a reproductive capacity, a self-referentiality in communication and in all the components of the communication system: structures, processes, boundaries, environments (Luhman 1988, see also p. 31 in Teubner 1993). We deviate, however, from Niklas Luhman's closed autonomous system and instead understand law as partially autonomous,[9] operating with, and dependent on—to various degrees—other systems operating in society (think of similar self-referential systems such as the world of science or that of government). Perceiving law, on the one hand, as not only conditioned by the social, and, on the other, as not only a fully autonomous, self-reproductive system, allows us to retain a productive tension between a biopolitical analytic on the one hand and that of legal form on the other.

Such an understanding of biolegality—we argue—allows for a more systematic analysis of what has been termed an "affirmative" approach to biopolitics. Early work on biolegality and biopolitical governance emphasized the continued forms of gendered, heteronormative, colonial, and racialized violence that come with the governance of biological material and people. Without dismissing the continued persistence of thanatopolitical and necropolitical dimensions of biopolitics (Mbembe 2008), biolegality also stresses the agency of law, its instruments, and its actors, emphasizing its potential to rearrange the biopolitical into an affirmative mode. For instance, the scholarship around biocitizenship (Petryna 2013; Rose and Novas 2005) details empirical and ethnographic ways in which biology and the biological self are figured as the procreant of political agency, prompting civic virtues to emerge in light of environmental disasters or biomedical events. These civic virtues—in times of anthropogenic ecological destruction—surpass those of humans. Calls to extend legal personhood to the more-than-human (animals, rivers, nature) are part of thinking biopolitics affirmatively (Rutherford and Rutherford 2013; Vatter and de Leeuw 2023).

Law and regulation are not only key facilitators to these biopolitical assertions but can themselves form the catalyst of an affirmative biopolitics. In other words, while early studies of biolegality have mostly covered ontic projects that focus on the implications of biopolitical modes of governmentality for the biological citizen, the asylum seeker, undocumented migrants, criminals, and legal persons in a variety of national contexts and transnational jurisdictions, our use of biolegality also includes ontological projects, those that focus on how new technologies of knowing and making life fundamentally affect how we understand life and law and what they might become.

A final note on our use of "legality" for it is this term that sets our approach apart from our companions in biopolitics, bioethics, and biolaw. The term, however, has several precedents, not all of which are useful to our approach. For instance, our use of legality does not include the idealist understanding proffered by a state. In other words, we are not necessarily interested in how (legal) authorities render a bioscientific research practice or biotechnological application "legal" or "illegal." Moreover, we also do not understand legality to be about the "property of being law"; i.e. as something that can only be enacted within formal legal institutions (see p. 7 in Shapiro 2011). Rather than defining legality in an idealist or materialist way, we focus instead on the emergent properties of legality, understanding legality as empirically derived and differentially constituted or enacted (see p. 335 in Silbey 2005, and 97 in Selznick 1961). Following the socio-legal scholar Susan Silbey, legality is an analytical term—not a sanctioned set of circumstances—that points to a variety of "meanings, sources of authority, and cultural practices that are recognized as legal, regardless of who employs them or for what ends" (see p. 353 in 2005). In the absence of the power of enforcement, it is the convention of social norms and traditions that guide and enforce legitimate and illegitimate behaviors—both in civic and scientific domains. Such an understanding of legality brings us to the core of our project, its engagement with knowledge and form.

Engaging Knowledge and Form

Drawing from the work of legal anthropology and legal geography, we understand legal forms to be cultural forms that stabilize over time (Maurer 2004; Strathern 2005; Ewick and Silbey 1998; Silbey 2005; Greenhouse 2014; Riles 2005). Whereas both law and science are seen as domains separated from society, the notion of coproduction—introduced by Jasanoff (2004)—allows us to see how law, science, and society are co-constitutive. Law—as a vital element of society—represents a core institution in the conditioning of science in society. Since regulations, rights, and legal governance form an integral part of this process, the work of law and that of its actors—lawyers, judges, and lawmakers—are paramount to capture the full breadth of the translation of science into society. Through the operation of coproduction we can decipher the different languages, methods, and interpretations that scientists and lawyers use to give meaning to developments in science. These approaches surpass the actual institutions of science and law and exist in our everyday lives and culture. Biolegality homes in on this coproduction between science and society, as well as science and law, to include the more specific effects that legalities have on the form and content of the biosocial subject or community.

We are cautious, however, not to deflate the term by suggesting its ubiquity. The legal anthropologist Bill Maurer remarked that socio-legal studies were experiencing a theoretical exhaustion like the one experienced by the social study of science, wherein law and science could now be discovered everywhere (2004, see also p. 477 in Silbey 2005). This exhaustion, he explains, has everything to do with the social construction paradigm that fails to give "any account of the network of human and nonhuman agents that, together, push back against the 'social' and in the process make their own moral claims known" (see p. 848 in 2004, see also Raffles 2001). Hence, rather than "law *and* culture," it should be "law *as* culture" just as science studies made the turn to "science *as* culture" (Latour 1999, 2012; Franklin 1995). In this shift, law and its instruments are not to be approached as a black box; instead, socio-legal scholars analyze what makes up the black box and describe what is happening empirically, ethnographically, and theoretically. We will show

how legal instruments comprise infrastructures, technologies, and labor that perform different functions and social effects in different contexts (Biagioli and Buning 2019; Cloatre 2018).

The acknowledgment of law *as* culture does not mean that we dismiss the autopoietic components of law (Luhmann 1988; Teubner 1993) in law's encounter with biology. On the contrary, our use of biolegality acknowledges that law operates as a distinct system that can autonomously deliberate and decide on how biological knowledge and technology are to be incorporated and reworked into legal knowledge. However, at the same time, we also acknowledge contributions from legal scholars working on symbolic-interaction theory (Bourdieu 1986; Van Klink et al. 2016) and media theory (Vismann 2013) who argue that the outside of law is always inside through its institutional culture and *praxis*. The understanding of law's institution as a self-referential system, while at the same time acknowledging its legal knowledges as a form of cultural technique, makes the analysis of biolegality possible.

The Australian legal scholar Margaret Davies aptly describes this process, namely that "[it] is neither the application of law to bioscience nor an account of the fundamental laws of biology, but rather the *generative and regenerative formations of both law and life*: the patterns, iterations, symbiotic relations, responses, and behaviors of life understood as law" (see p. 4 in 2020, our emphasis). Put in other words, configurations of law and biology create *biolegalities*: a structure of legality that allows the formation of new epistemologies as well as new ontologies of biology, nature, life, materiality, and sociality. Law is therefore not only another form of culture, nor a mere extension of state politics; there's a distinctiveness to law that features prominently in contemporary reconfigurations of biology and nature. In this sense—namely in the interaction of the changing understanding of the laws of life (i.e. bioscientific laws) and the iterative, self-reflexive, and recombinant process of juridical law making—a theory of biolegalities intervenes in the dichotomy between a rigorous legal pluralism and a strict conception of sovereign law or legal governance (Cotterell 1992). While it recognizes the social construction of law, biolegalities is premised upon the recognition that the law—with its unique legal forms, doctrines, and techniques—has a distinctive quality and action. In suggesting that biolegal arrangements trouble or challenge existing legal forms, our approach yields an understanding of biolegal arrangements as bearing on both challenges of an epistemological and ontological order. By this we mean that knowledge or being transform alongside the new premises of biolegality. These transformations do not always appear in concert; their changes demonstrate in fact the incommensurability of legality vis-à-vis epistemic or ontological conditions.

Anthropos in the Age of Biolegality

> The issues of what things are, which new things are significant, how they are significant, and how one knows they are new and significant are of prime importance pragmatically, anthropologically, and philosophically.
> (Paul Rabinow and Gaymon Bennett, *Designing Human Practices*)

The careful reader of these introductory pages will notice a slippage between the use of biolegality as a conceptual or heuristic device, and the use of biolegality as a historical or empirical condition. The slippage is intentional and mirrors the two main aims of our project. So far, we have mostly engaged biolegality as a heuristic device that makes the constitutive effects of law and bio(techno)logy legible. But using biolegality to describe a historical process or condition is equally salient. While this historical process is discussed at length in part I (Chapters 2 and 3), let us briefly identify three recent historical developments that inform the rearrangement of law and life in the twenty-first century: (1) the economization of science and law through globalized regimes of privatization, (2) the posthuman implications of biotechnology, and (3) the pressures caused by anthropogenic climate change.

First, as many scholars observe, it was not only the scientific ability to splice genes in the 1970s that brought biotechnology to what it is today; the commercialization of the life sciences fundamentally changed the way research was organized. In many places, the biosciences moved from public research institutions into private firms, which utilize basic science for commercial applications, necessitating the revisions and extension of intellectual property regimes. The "reinvention of life" became one of the most profitable industries of the twenty-first century (Parry 2004: 3-4). This economic development could not have taken place without the facilitation of law, demonstrated most poignantly by the 1980 *Diamond v. Chakrabarty* case in the US. Here, the Supreme Court ruled that the Indian-born Ananda Chakrabarty, who worked for General Electric, could patent a modified bacterium that breaks down crude oil. The ruling significantly expanded the privatization of science by shifting relations between scientists, corporations, and research institutions, allowing scientists, companies, and universities to patent just about anything under the sun, enabling the commercialization of biological material across the world (Mirowksi 2011; Callon 1994). As we will describe in more detail in Chapter 3, which examines the debate around the commercialization of human genetic material, these developments impacted legal form as well as scientific and commercial practice. Most acutely we witness how turning human genetic material into commodifiable entities troubles the modern legal division between persons and things.

The commercial developments of the life sciences, enabled via law and global governance systems, have not only changed how corporations or universities appropriate biological material, but also how individuals and communities relate to their own biological and genetic material—both in terms of monetary value as well as the knowledge and meanings attached to them. Sheila Jasanoff's work on bioconstitutionalism (2011) is compelling in this respect. She and others outline the constitutive effects that the life sciences, in coproduction with legal conceptions of life, have on the "*is* and *ought* in modern times" (see p. 3 in Ibid.). Alluding to the genetic age, she describes how constitutional systems struggle to "address and accommodate new claims and questions by citizens empowered with genetic knowledge" (see p. 94 in ibid.). Biological knowledge can aid citizens' quest to reorder their democratic dispositions vis-a-vis the nation-state.

While the instances of biolegality demonstrated in the next few chapters suggest similar operations of bioconstitutionalism, we also suggest that developments in what has been termed the "postgenomic" age are complicating genetic knowledge and therefore challenging the anthropocentric grounds of bioconsitutionalism driving analysis beyond the subject–state nexus. As we will describe, postgenomic knowledge characterizes itself by the destabilization of the unitary body through the body's entanglement with the environment (Niewöhner 2011; Meloni 2015; Van Wichelen and Keaney 2022) and by the digitalization of life where the self becomes disaggregated through algorithms and the integration of big data (Richardson and Stevens 2015; Leonelli and Ankeny 2012). These characteristics allow for more-than-human investigations of biolegality, those that transcend the frame of national constitution, and move analysis beyond the realm of discourse into material praxis and materialities (Latour 2013; Pottage and Sherman 2010; Cloatre 2018; Kang 2018; Murray 2019; Kang and Kendall 2019).

These diverse examinations of biolegality in the contemporary context confirm that technoscientific developments in the realm of biology are themselves fundamentally testing the modern distinction between persons and things, and humans and non-humans. For instance, medical technologies such as transplants and the development of artificial (bio)materials for therapeutic ends have pushed the boundaries of personhood and humanness. In her work on pig-human embryos, the legal scholar Yvonne Cripps (2018) explains, how gene editing makes possible the formation of transspecies organs that trouble not only the idea of ownership but also the premises on which legal personhood is defined. Targeted and applied within a promissory context of precision medicine, the advent of embryonic human–non-human chimeras combined with the CRISPR-Cas9 technique may redefine what it means to be human depending on whether some of these chimeric forms—despite prohibitions—get to become patented. Moreover, new insights from immunology, research on the microbiome, and epigenetics have further complicated our thinking about human boundedness and exceptionalism. Novel understandings from microbiomic science, for example, show that the human body is dependent on the many microbiotas in and on our bodies. While scholars in the social study of science have brought these insights to bear on posthuman, multi-species, and more-than-human theorization (Helmreich 2015; Tsing 2015), legal scholars are yet to study the implications of these postgenomic scientific developments on the legal understanding of the human and the person.

Third, this problematic is further heightened in the age of the Anthropocene, where the burden humans have put on natural resources, land, soil, water, air, and non-human life, including plants and microbes, is destroying the natural world. Hence, the more-than-human context of biolegality also includes how new biotechnologies—such as gene editing and gene drives—are affecting the environment and specific ecologies (Braverman 2017). Gene drives—a faster—and multiplying—way of spreading genetic variants through a population than traditional Mendelian inheritance—can alter the genetic structure of an entire species.[10] Since ecological systems—like climate change—do not stop at national borders, the global governance of such new technologies is of utmost importance. What scholars in these novel biotechnological contexts who address more-than-human dimensions of emerging technologies have in common is that they do not draw on bioethics or biolaw, which they consider anthropocentric (Braverman 2017; Franklin 2006; Lorimer 2020; Asdal et al. 2016). Instead, they draw from philosophy, anthropology, geography, and more specialized fields such as science and technology studies (STS) or environmental humanities to inform their approach. Biolegality offers an apt description of a time and place in which these more-than-human dimensions of life and law can be articulated.

Our study of biolegality sits across these historical and conceptual developments. By consolidating inquiries into the matter of the human or the *Anthropos,* we engage the economization of science on the one hand, and the more-than-human on the other. Biolegality, as a spatiotemporal condition, pushes the boundaries of humanness and humanist reason. While novel developments in biomedicine and biotechnology define new ecologies of human individuality, law—in responding to these developments—correspondingly redefines the *anthropos*. Biolegal arenas, thereby, present new understandings of humans, human relations, and human societies. Such posthuman or more-than-human understandings are pivotal to articulating new conceptual tools in an age of anthropogenic destruction. Planetary perspectives—those that recognize the Anthropocene thesis and adopt ecological approaches in an attempt to fundamentally reconfigure humans' relationship with nature—rely on biolegal analysis to formulate and achieve effective planetary stewardship for a sustainable life on earth.

Our critical intervention, then, is as modest as it is ambitious. On the one hand, our goal is to provide an advanced introduction and simply bring together recent scholarship from legal theory and science studies that tackle the coproduction of law and life, illustrating the usefulness of biolegality as a heuristic device complementing bioethics and biolaw. We hope this will give non-specialist readers and students interested in this topic, an initiation to the field. On the other hand, our intervention also argues that biolegal developments are pushing the boundaries of the *anthropos* in Western modern knowledge. The constitutive relation between law and biology today is reconfiguring what we understand as *the human*. The biolegal synthesis prompts new forms of being in the world, generating ontological effects. While shifts of these kinds are traditionally debated within the realm of philosophy or anthropology, law and legality are in fact privileged sites for action and meaning. Biolegality is at the heart of these sites.

Outline of the Book

Our programmatic and critical stance is that biolegality helps illuminate how the biosciences and law are reconfiguring the *anthropos* and her relations. To place this discussion in a longer historical narrative about the relationship between biology and law, we start, in part I, with the examination how law has long tried to appropriate, control, or manage, first nature, then biology and more recently, biotechnology. Hence, Chapter 2, "From Nature to Biology," provides a brief genealogy of the changes law has undergone through its confrontation with nature and biology. We historicize key debates around law and nature, and engage with scholarship emerging from the history of natural law, evolutionary jurisprudence, sociology of law, philosophy of law, feminist jurisprudence, and legal pluralism. We describe the persistence of economic models in the current legal field and situate the currency of these models within a modern and postmodern framework. This opens the context for our intervention in Chapter 3, "The Biotechnology Revolution," where we examine the extent to which new arrangements of law and legal form—borne out of interactions with biotechnology—challenge their foundational premises. With the help of anthropology and science and technology studies, we demonstrate two biolegal directions: one in which law usurps biotechnological changes, and one in which biotechnology challenges legal form in ways that are still open and undetermined.

The four chapters in part II encompass core investigations of biolegal shifts, focusing on developments that trouble the *anthropos*. Hence, in our respective chapters on "Genes," "Brains," "Babies," and "Bodies," we analyze biolegal changes that are reshaping knowledge forms attached to the human in western modernity by tracing the interaction between specific biotechnological advances and legal practices. As we propose, these biolegal shifts influence how the human, her relations and environments in modernity are reconfigured, namely through (1) exchange and property, (2) self and personhood, (3) kinship and parenthood, and (4) community and identity.

In Chapter 4 on "Exchange and Property," we examine how co-productive relationships between biotechnology and intellectual property rights redefine the course of discovery and invention. Focusing on human genes, we describe how isolating and splicing genes from bodies on the one hand, and commercializing biological material on the other, complicates human attachments to their biological materials. Blood, tissue, semen, organs, breastmilk, placenta, and stool were often seen—and legally regarded—as waste. Genomic, reproductive, and transplant technologies (organ, reproductive or fecal) have altered the value and meaning of these biological materials, thereby confounding how they should be governed. Our chapter tackles these complications around the governance of genetic material by focusing on the central problem of how law defines genetic materials and how this aligns with people's and society's understanding of what human genetic material entails and represents. High-profiled court cases such as the *Moore v. Regents of the University of California*, where John Moore sued UCLA for using "his" tissue to develop cell lines that were further commercialized, reveal the discrepancies between how individuals relate to their bodies and "their" biological materials, and how institutions—such as universities or pharmaceutical companies—use these materials for research and commercial purposes. By engaging the anthropological problem of exchange, we explore property law's relation to nature from the perspective of biolegality.

In "Self and Personhood," the subject of Chapter 5, we examine the new boundaries produced by the legal puzzles emanating from the encounter between legal personhood and recent neuroscientific developments. The latest discoveries in the neurosciences are starting to trouble the notion of "personality" and, consequently, the social problem of responsibility. MRI scans are increasingly used in courts to verify whether particular behaviors were intentional, putting into question who committed certain acts or how responsibility for these acts are adjudicated to a particular person. As such the MRI scan might travel the same route as "DNA" by being reworked, after its introduction to the criminal process, as a legally applicable form of evidence. The legal brain, then, becomes a prime field in which legal and biological knowledge simultaneously challenge and produce each other's validity. This chapter demonstrates how new practices and understandings of neuroscience are affecting social, cultural, and legal conceptions of personhood and how different approaches to the meaning of law engender particular configurations of biolegality.

Chapter 6 on "Kinship and Parenthood" engages the field of reproductive technologies. Developments in this domain are changing biological and social relations and altering the concept of legal parenthood. The conventional biological arrangement that relies on the gestating mother and the recognition of legal paternity by the father is being replaced by in vitro conceptions, genetic fatherhood, mitochondrial replacement techniques, gestational surrogacy, and embryo adoption. Because it is now possible to engineer reproduction, parenthood becomes fractionalized, making it possible to have many forms of biological and legal parenthood: genetic, gestational, mitochondrial, intentional. Nonetheless, many contemporary cases reveal the extent to which the legal system struggles to integrate these new arrangements with the existing normative frameworks for paternity and reproduction. As such, we investigate the mutating constructions of kinship, maternity, and paternity in courts as they work through scientific developments and global processes.

"Community and Identity"—Chapter 7—analyzes the shifting biolegalities of communities as: (1) collectivization of biological suspects in forensic and immigration practice; (2) the making of biological populations for health justice; and (3) the biological consolidation of identity and rights as forms of citizenship and nation building. First, the use of standardized "control populations" in forensics to establish DNA profiles for suspects has reversed the person's (or entire community's) "right to remain silent" because the "silent witness" of DNA always speaks. The enormous increase in population databases—synthesizing human DNA with digital data—makes law the prime location where the biolegality of an entire community (including the nation) can be solidified or contested. Second, the emergence of personalized, community, or race-based medicine constructs new communities based on biological markers. While some practices rely on older constructions of race, ethnicity, or community, we demonstrate how science and publics draw new boundaries based on genetic, epigenetic, microbiomic, or neural distinctions that, in turn, produce new social identities and rights. Third, we show how genomic and postgenomic insights contribute to new ways in which communities and citizens present themselves to each other. As we will argue, in all these cases, biology contributes to determining one's (legal) belonging and reconfigures one's relation toward rights, civic identity, community, and the nation-state.

Finally, the coda to the book engages with the core problematics of biolegality by reflecting on the state of law and biology during and after the global COVID-19 pandemic through mathematic disease modeling, which prompted the enactment of a suite of legalities. While sustaining the idea of the technological fix, the focus on epidemic intelligence in the age of information and postgenomics aims at more than merely the control of COVID-19; mathematic disease modeling has applications beyond the immediate emergency response to one pandemic or pandemics generally. While modeling produces new frameworks of expertise and public health sovereignty (Jasanoff 2020), it also functions as a crisis technology (Anderson 2021) prompting reductionist solutions focused on containment in lieu of more ecological models of disease control. Our aim in this chapter is to explore how the state, law, and legality are implicated in these various models and what it might mean for law if our hope for a more sustainable future lies in fostering the planetary rather than global approaches to health and biomedicine. The biolegal synthesis—we propose—generates important questions essential to planetary futures.

Notes

1. There was and still is great contestation around who should get the credit for first inventing the gene editing tool. Others, among them Virginijus Šikšnys from Vilnius University and Feng Zhang from the Broad Institute, had published similar findings around the same time. See Greely (p. 121 in 2019) for a discussion of the patent politics surrounding the events. In the end, it was Doudna and Charpentier who received the Nobel Prize for Chemistry for CRISPR-Cas9 technique, see https://www.nytimes.com/2020/10/07/science/nobel-prize-chemistry-crispr.html.
2. See https://www.nytimes.com/2018/11/26/health/gene-editing-babies-china.html. See Greely (2019) for a critical discussion on He's justifications around the medical necessity of his actions and Hurlbut (2020) for an alternative, though no less critical, view around narratives of "medical necessity" that go beyond the biological.
3. Until the He Jiankui case, clinical trials in the US and Europe using gene editing focused exclusively on somatic cells and did not include genetic modification of germlines, which involves reproductive cells such as sperm, eggs, zygotes, and embryos. Gene editing on somatic cells takes place on the level of the individual while germline modifications encompass the genetic makeup of future generations.
4. There is an important distinction to be made between the editing of somatic cells that can take place anywhere in the body and the editing of germ cells that encompass heritable reproductive cells. With the case of the Chinese twins, somatic cells in germ cells were targeted.

5. See for a general overview of biopolitics Lemke 2010, 2011; Mills 2017, 2017; and Pugliese 2012, for an account of biopolitics and biotechnology see Obasogie and Darnovsky 2018, for biomedicine see Rose 2001, 2007, and reproduction see Waldby and Cooper 2008, which has been further refined by concepts such as biocapital (Cooper 2011; Rajan 2006), biolegitimacy (Fassin 2009, 2012), and biocitizenship (Heinemann & Lemke 2014, Petryna 2004; Rose and Novas 2005).
6. For a similar argument, see also Sarah Franklin, 'Life itself: global nature and the genetic imaginary', in Sarah Franklin, Celia Lury and Jackie Stacey, *Global Nature, Global Culture* (London: Sage 2000), 188/227; Sarah Franklin and Margaret Lock (eds), *Remaking Life and Death: Toward an Anthropology of the Biosciences* (Santa Fe, NM: School of American Research/Oxford: James Currey 2003).
7. For an excellent analysis of biolegal technologies outside of forensics see Arum Budiastuti (2017).
8. Though some countries demonstrate an ambivalent balance between the securing of human rights and the prevention of fraud, DNA testing still operates as an "arbiter of truth" (see p. 76 in Hall and Naue 2014). Moreover, in the realms of immigration, informed consent and professional ethics are less prominent than in medical practice. These issues all point to a more complex understanding of biological citizenship. Whereas the focus on the term has thus far been "on the extension of rights, the emergence of new possibilities of civic participation and social engagement, and on the choice-enhancing options of biomedical technologies" (see p. 104 in Heinemann and Lemke 2014), Heinemann and Lemke (see p. 105 in ibid.) argue that the scholarship overlooks "practices of surveillance and exclusion and the refusal of citizenship rights based on biological knowledge."
9. This resonates more with Gunther Teubner's work, who argues that "[w]e may conclude by saying that a legal system, becomes autonomous to the extent that it manages to constitute its components—action, norm, process, identity—into self-referential cycles. It achieves autopoietic autonomy only when the

components of the system formed in this way are linked together to form a hypercycle" (see p. 27 in Teubner 1993).

10. One of the conceivable applications currently being considered, for example, is that of altering the genetic makeup of mosquito populations in such a way that they are not receptive anymore to carrying vector-born human pathogenic diseases such as malaria, dengue, zika, or yellow fever. Besides the vast benefits that these developments might have for global health, the engineering of whole species populations could also have substantial consequences for the extinction of certain species and/or unintended effects on species that were not targeted (Esvelt 2017; Hefferon and Herring 2020).

References

Andorno, R., Baylis, F., Darnovsky, M., Dickenson, D., Haker, H., Hasson, K., … and Zuscinova, J. 2020. Geneva Statement on Heritable Human Genome Editing: The Need for Course Correction. *Trends in Biotechnology*, 38(4), pp. 351–354.

Anderson, W. 2021. The Model Crisis, or How to Have Critical Promiscuity in the Time of Covid-19. *Social Studies of Science*, 51(2), pp. 167–188.

Asdal, K., Druglitro, T., and Hinchliffe, S., eds. 2016. *Humans, Animals and Biopolitics: The More-Than-Human Condition.* Routledge.

Barrangou, R. and Doudna, J. A. 2016. Applications of CRISPR Technologies in Research and Beyond. *Nature Biotechnology*, 34(9), pp. 933–941.

Biagioli, M. and Buning, M. 2019. Technologies of the Law/ Law as a Technology. *History of Science*, 57(1), pp. 3–17.

Bourdieu, P. 1986. The Force of Law: Toward a Sociology of the Juridical Field. *Hastings Law Journal*, 38, p. 805.

Braverman, I., ed. 2017. *Gene Editing, Law, and the Environment: Life Beyond the Human.* London: Routledge.

Brown, W. 2002. *Moralism as Antipolitics Materializing Democracy: Toward a Revitalized Cultural Politics*, pp. 368–392. London: Duke University Press.

Budiastuti, A. 2017. In DNA We Trust? Biolegal Governmentality and Illegal Logging in Contemporary Indonesia. *East Asian Science, Technology and Society: An International Journal*, 11(1), pp. 51–70.

Callon, M. 1994. Is Science a Public Good? Fifth Mullins Lecture, Virginia Polytechnic Institute, March 23, 1993. *Science, Technology, & Human Values*, 19(4), pp. 395–424.

Cooper, M. E. 2011. *Life as Surplus: Biotechnology and Capitalism in the Neoliberal Era*. Washington: University of Washington Press.

Cotterrell, R.B. 1992. *The Politics of Jurisprudence: A Critical Introduction to Legal Philosophy*. University of Pennsylvania Press.

Cloatre, E. 2018. Law and ANT (and Its Kin): Possibilities, Challenges, and Ways Forward. *Journal of Law and Society*, 45(4), pp. 646–663.

Cripps, Y. 2018. The Global Person: Pig-Human Embryos, Personhood, and Precision Medicine. *Indiana Journal of Global Legal Studies*, 25(2), pp. 701–727.

Davies, M. 2020. Distributed Cognition, Distributed Being, and the Foundations of Law. In *Personhood in the Age of Biolegality*, pp. 205–223. Cham: Palgrave Macmillan.

Doudna, J. A. and Steinberg, S. H. 2017. A Crack in Creation: Gene Editing and the Unthinkable Power to Control Evolution. *The FASEB Journal*, 32(1), pp. 3–4.

Ewick, P. and Silbey, S. S. 1998. *The Common Place of Law: Stories from Everyday Life*. Chicago: University of Chicago Press.

Esvelt, K. M. 2017. Rules for Sculpting Ecosystems: Gene Drives and Responsive Science. In *Gene Editing, Law, and the Environment*, pp. 20–37. London: Routledge.

Fassin, D. 2009. Another Politics of Life Is Possible. *Theory, Culture & Society*, 26(5), pp. 44–60.

Foucault, M. 1976. *The Will to Knowledge: The History of Sexuality*, vol. 1 (trans. R. Hurley, 1998). London: Penguin.

Franklin, S. 1995. Science as Culture, Cultures of Science. *Annual Review of Anthropology*, 24(1), pp. 163–184.

Franklin, S. 2006. The Cyborg Embryo: Our Path to Transbiology. *Theory, Culture, and Society*, 23(7–8), pp. 167–187.

Greenhouse, C. J. 2014. Unexpected Properties: Strathern on the Relation of Law and Culture. *Theory, Culture & Society*, 31(2–3), pp. 167–184.

Greely, H. T. 2019. CRISPR'd Babies: Human Germline Genome Editing in the 'He Jiankui Affair'. *Journal of Law and the Biosciences*, 6(1), pp. 111–183.

Hacking, I. 2006. Genetics, Biosocial Groups & The Future of Identity. *Daedalus*, 135(4), pp. 81–95.

Heinemann, T. and Lemke, T. 2014. Biological Citizenship Reconsidered: The Use of DNA Analysis by Immigration Authorities in Germany. *Science, Technology, & Human Values*, 39(4), pp. 488–510.

Hefferon, K. and Herring, R. 2020. Engineered Gene Drives and Their Value in the Control of Vector-Borne Diseases, Weeds, Pests, and Invasive Species. In *GMOs*, pp. 401–419. Cham: Springer.

Hurlbut, J. B. 2020. Imperatives of Governance: Human Genome Editing and the Problem of Progress. *Perspectives in Biology and Medicine*, 63(1), pp. 177–194.

Helmreich, S. 2015. *Sounding the Limits of Life: Essays in the Anthropology of Biology and Beyond*. Princeton: Princeton University Press.

Jasanoff, S. 1997. *Science at the Bar: Law, Science, and Technology in America*, vol. 9. Cambridge: Harvard University Press.

Jasanoff, S. 2001. Ordering Life: Law and the Normalization of Biotechnology. *Politeia*, 62, pp. 34–50.

Jasanoff, S., ed. 2004. *States of Knowledge: The Co-production of Science and the Social Order*. London: Routledge.

Jasanoff, S., ed. 2011. *Reframing Rights: Bioconstitutionalism in the Genetic Age*. Cambridge: MIT Press.

Jasanoff, S. 2020. Pathologies of Liberty: Public Health Sovereignty and the Political Subject in the Covid-19 Crisis. *Cahiers Droit, Sciences & Technologies*, 11, pp. 125–149.

Jasanoff, S., Hurlbut, J.B., and Saha, K. 2015. CRISPR Democracy: Gene Editing and the Need for Inclusive Deliberation. *Issues in Science and Technology*, 32(1), pp. 25–32.

Kang, H.Y., 2018. Law's Materiality: Between Concrete Matters and Abstract Forms, or How Matter Becomes Material. In *Routledge Handbook of Law and Theory*, pp. 453–474. Routledge.

Kang, H.Y. and Kendall, S. 2019. Legal Materiality. In S. Stern, M. DelMar, and B. Meyler (eds.), *The Oxford Handbook of Law and Humanities*, pp. 21–38. Oxford, UK: Oxford University Press.

Latour, B. 1999. *Pandora's Hope: Essays on the Reality of Science Studies*. Cambridge: Harvard University Press.

Latour, B. 2012. *We Have Never Been Modern*. Cambridge: Harvard University Press.

Latour, B. 2013. *An Inquiry into Modes of Existence*. Harvard University Press.

Lemke, T. 2010. From State Biology to the Government of Life: Historical Dimensions and Contemporary Perspectives of 'Biopolitics'. *Journal of Classical Sociology*, 10(4), pp. 421–438.

Lemke, T. 2011. Biopolitics. In *Biopolitics*. New York University Press.
Leonelli, S. and Ankeny, R.A. 2012. Re-thinking Organisms: The Impact of Databases on Model Organism Biology. *Studies in History and Philosophy of Biological and Biomedical Sciences* 43(1), 29–36.
Lezaun, J. 2006. Creating a New Object of Government: Making Genetically Modified Organisms Traceable. *Social Studies of Science*, 36(4), pp. 499–531.
Luhmann, N. 1988. Law as a Social System. *Northwestern University Law Review*, *83*, p. 136.
Lorimer, J. 2020. *The Probiotic Planet: Using Life to Manage Life*. Minneapolis: University of Minnesota Press.
Lynch, M. 2003. God's Signature: DNA Profiling, the New Gold Standard in Forensic Science. *Endeavour*, 27(2), pp. 93–97.
Lynch, M. and McNally, R. 2009. Forensic DNA Databases and Biolegality. *Handbook of Genetics and Society*, pp. 283–301. London: Routledge.
Lynch, M., Cole, S. A., McNally, R., and Jordan, K. 2010. *Truth Machine: The Contentious History of DNA Fingerprinting*. Chicago: University of Chicago Press.
Maurer, B. 2004. The Cultural Power of Law? Conjunctive Readings. *Law & Society Review*, 38(4), pp. 843–850.
Mbembe, A. 2008. Necropolitics. In *Foucault in an Age of Terror: Essays on Biopolitics and the Defence of Society*, pp. 152–182. London: Palgrave Macmillan UK.
M'charek, A., Schramm, K., and Skinner, D. 2014. Topologies of Race: Doing Territory, Population and Identity in Europe. *Science, Technology, & Human Values*, 39(4), 468–487.
Meloni, M. 2015. Epigenetics for the Social Sciences: Justice, Embodiment and Inheritance in the Postgenomic Age. *New Genetics and Society*, 34(2), pp. 125–151.
Mills, C. 2017. *Biopolitics*. London: Routledge.
Mirowski, P. 2011. *Science-mart: Privatizing American Science*. Harvard University Press.
Niewöhner, J. 2011. Epigenetics: Embedded Bodies and the Molecularisation of Biography and Milieu. *BioSocieties*, 6(3), pp. 279–298.
Obasogie, O. K. and Darnovsky, M., eds. 2018. *Beyond Bioethics: Toward a New Biopolitics*. Berkeley: University of California Press.
Parry, Bronwyn. 2004. *Trading the Genome: Investigating the Commodification of Bio-information*. New York: Columbia University Press.
Petersen, A. 2011. *The Politics of Bioethics*. Routledge.

Petryna, A. 2004. Biological Citizenship: The Science and Politics of Chernobyl-exposed Populations. *Osiris*, 19, pp. 250–265.
Petryna, A. 2013. *Life Exposed: Biological Citizens after Chernobyl*. Princeton University Press.
Poland, S. C. 2005. Bioethics, Biolaw, and Western Legal Heritage. *Kennedy Institute of Ethics Journal*, 15(2), pp. 211–218.
Pottage, A. 2007. The Socio-legal Implications of the New Biotechnologies. *Annual Review of Law and Social Science*, 3, pp. 321–344.
Pottage, A. and Sherman, B. 2010. *Figures of Invention: A History of Modern Patent Law*. Oxford University Press.
Pugliese, J. 2012. *Biometrics: Bodies, Technologies, Biopolitics*. London: Routledge.
Rabinow. P. 1992. Artificiality and the Enlightenment: From Sociobiology to Biosociality. In Crary, Jonathan and Kwinter, Sanford (eds.), *Incorporations*. New York: Zone Books.
Rabinow, P. 2010. Artificiality and Enlightenment: From Sociobiology to Biosociality. *Politix*, 2, pp. 21–46.
Raffles, H. 2001. The Uses of Butterflies. *American Ethnologist*, 28(3), pp. 513–548.
Rabinow, P. and Bennett, G. 2012. *Designing Human Practices: An Experiment with Synthetic Biology*. University of Chicago Press.
Rajan, K. S. 2006. *Biocapital: The Constitution of Postgenomic Life*. Durham: Duke University Press.
Richardson, S.S. and Stevens, H., eds. 2015. *Postgenomics: Perspectives on Biology after the Genome*. Duke University Press.
Riles, A. 2005. A New Agenda for the Cultural Study of Law: Taking on the Technicalities. *Buffalo Law Review*, 53, p. 973.
Rose, N. 2001. The Politics of Life Itself. *Theory, Culture & Society*, 18(6), pp. 1–30.
Rose, N. and Novas, C. 2005. Biological Citizenship. *Global Assemblages: Technology, Politics, and Ethics as Anthropological Problems*, pp. 439–463. Oxford: Blackwell Publishing.
Rose, N. 2007. Molecular Biopolitics, Somatic Ethics and the Spirit of Biocapital. *Social Theory & Health*, 5(1), pp. 3–29.
Rutherford, S. and Rutherford, P. 2013. Geography and Biopolitics. *Geography Compass*, 7(6), pp. 423–434.
Silbey, S.S. 2005. Everyday Life and the Constitution of Legality. In Jacobs, M. and Hanrahan, N. (eds.), *The Blackwell Companion to the Sociology of Culture*, pp. 332–345. Oxford: Blackwell.

Shapiro, S. J. 2011. *Legality*. Cambridge: Harvard University Press.

Selznick, P. 1961. Sociology and Natural Law. *Natural Law Forum*, 6, pp. 84–108.

Sparrow, R. and Mills, C. 2021. Genome Editing: From Bioethics to Biopolitics. *BioSocieties*, 17(3), pp. 485–505.

Strathern, M. 1992. Parts and Wholes: Refiguring Relationships in a Post-Plural World. In *Adam Kuper Conceptualizing Society*, pp. 75–106. London: Routledge.

Strathern, M. 2005. *Kinship, Law and the Unexpected: Relatives Are Always a Surprise*. Cambridge: Cambridge University Press.

Swiffen, A. 2010. *Law, Ethics and the Biopolitical*. Routledge.

Teubner, G. 1993. *Law as an Autopoietic System*. Oxford and Cambridge: Blackwell Publishers.

Tsing, A.L. 2005. *Friction: An Ethnography of Global Connection*. Princeton University Press.

Tsing, A. L. 2015. *The Mushroom at the End of the World*. Princeton: Princeton University Press.

Van Beers, B.C. 2020. Rewriting the Human Genome, Rewriting Human Rights Law? Human Rights, Human Dignity, and Human Germline Modification in the CRISPR Era. *Journal of Law and the Biosciences*, 7(1), pp. 1–36.

Van Klink, B., Van Beers, B., and Poort, L. eds. 2016. *Symbolic Legislation Theory and Developments in Biolaw*. Springer International Publishing.

Van Wichelen, S. and Keaney, J. 2022. The Reproductive Bodies of Postgenomics. *Science, Technology, & Human Values*, 47(6), pp. 1111–1130.

Vatter, Miguel and De Leeuw, Marc. 2023. Human Rights, Legal Personhood and the Impersonality of Embodied Life. *Law, Culture and the Humanities*, 19(1), pp. 106–125.

Vismann, C. 2013. Cultural Techniques and Sovereignty. *Theory, Culture & Society*, 30(6), 83–93.

Winickoff, D. E. 2015. 2 Biology Denatured: The Public-Lives of Lively Things. In Hilgartner, S., Miller, C. A., and Hagendijk, R. (eds.), *Science and Democracy: Making Knowledge and Making Power in the Biosciences and Beyond*, p. 15. London: Routledge.

Waldby, C. and Cooper, M. 2008. The Biopolitics of Reproduction: Post-Fordist Biotechnology and Women's Clinical Labour. *Australian Feminist Studies*, 23(55), pp. 57–73.

Part I

Law and Life

2

From Nature to Biology

Introduction

Beginning with classical philosophy, the fundamental binary of Western thought is the dualism between *Physis* and *Nomos*: the natural versus the cultural world. Here, the animal stands opposed to man, one representing an order imbued by a violent and instinctive nature, the other by values and culture. Ever since, society has preoccupied itself with the question of whether *Nomos* has its own "natural" order: do humans and human culture also obey certain universal natural laws? Is our striving for the good and a good life an innate quality (Aristotle)? Do our moral beliefs emerge from our passions (Hume), or our ability to reason (Kant), or just from a need for survival (Hobbes)? Is the social contract (Rousseau) an attempt to institutionalize empathy and property claims (Locke), a fair distribution of material goods and opportunities (Rawls), an equal communicative recognition and access (Habermas); or are economic structures "naturally" destructive (Marx), or "naturally" liberating (Hayek)? Does human nature have an inherent demand for sociality, liberty, and equality?

This set of questions is what preoccupies natural law, and in this chapter, we trace some of the key transitions natural law has undergone

S. van Wichelen and M. de Leeuw, *Biolegality*, Biolegalities,
https://doi.org/10.1007/978-981-99-8749-8_1

through its confrontation with the changing definitions, categorizations, and understanding of nature, and—since the nineteenth century—biology. It is not our intention to rehearse the history of natural law here, for instance by approaching it as a predecessor to the contemporary relation between law and biology. Instead, this chapter provides a conceptual overview of how two fields of order—law and biology—reaches a form of theoretical reciprocity or space of "liminality." It does so by analyzing the uses and abuses of nature and their consequences for a moral, political, and natural understanding of ourselves. In this sense, natural law can be understood as a hybrid concept, where a *naturalized* law and a *legalized* nature are products of a history of "boundary work" (Gieryn 1983) in the terrains of law and science. We are thus concerned with how nature and biology determine what law and legality can be and do, and, vice versa, how law and legality determine what nature/biology can be and do. Inspired by contemporary theoretical perspectives in STS (Lynch et al. 2010; Jasanoff 1997, 2001) and the central work of legal scholars of nature and science (Braverman 2017; Delaney 2003; Kelley 1990; Pottage 2007), we describe in the chapters of Part II some instances where nature, biology, law, and legality intertwine, how ideas in these domains circulate (or not) and co-produce new social forms.

Rather than assuming that law derives its normative claims from those about nature, the more interesting question for us is how thinking about nature, and thinking about law, results in a specific dialectic of law and nature/biology. It is this dynamic that produces distinctive and shifting knowledge about life and law, producing new scientific and legal practices. A simple example is our notion of the brain. In early Greek philosophy the brain was seen as "the seat of understanding" and the location of the mind (Huffman 2021). This biological presupposition (we *are* our brain) created a normative presupposition (if I know your brain, I know who you are and why you act the way you do), still echoes through in contemporary neurolaw debates despite having undergone many radical changes following new insights and advanced tools like the EEG (electroencephalogram) or MRI-scans. The growing belief in neuroscience impacted what kind of treatments were permitted by law: a frontal lobotomy, direct cutting into the brain tissue via the eye, or electro-shock treatment to "re-set" the neural pathways were legally allowed procedures. Current brain interventions mediated by medical "enhancement" pills to relief depression or pain are legally accessible although sometimes have severe side effects (Prozac), or similar opiate levels as illegal drugs (OxyContin). In all these examples, from brain shocking to pain prescription, a particular relation between scientific and legal knowledge determines if a medical practice is allowed or not, and what kind of suffering or pain is considered real or legitimate. It illustrates the way normative conceptions of the brain since antiquity continue to dominate contemporary deployments of neuroscience in legal contexts, namely that neurolaw reproduces the presuppositions of the mind body dualism.

This chapter starts off from the ancient Greeks leading to contemporary tenets of natural law thinking, proffered by philosophers and legal scholars alike, and premised on basic human needs and values. If all humans are perceived as intrinsically inclined to do good, law has the task to make this quality explicit by making good and just practices and actions apparent to the world. Or, in contrast, if humans are perceived as intrinsically self-interested and potentially evil, it justifies law as an instrument for disciplining our unreliable nature. While the dominance of legal positivism and American pragmatism in law disconnects law from moral and political questions (focusing the study of law on its internal structures and procedural legitimations), Marxist and feminist approaches re-contextualized law toward social and economic questions (focusing on the study of class, patriarchy, power relations, and state authority). An important third "stream" entering the legal debate since the 1970s and 1980s concerns the study of human biology centering on human agency through sociobiology, cognitive psychology, and neuroscience, and offering a natural scientific dimension to law and criminology. It is through the lens of economics that legal positivism and pragmatism can be aligned with evolutionary biology—an important relation discussed toward the end of the chapter.

From Nature to Biology: Philosophical Foundations

The natural tradition insists that what can be learned from our nature should be applied and implemented via law and the political system. However, the meaning of nature—and thus *human* nature—largely depends on the philosophical, religious, political, economic, and scientific worldview through which nature is examined.

According to the prominent legal philosopher John Finnis, the most basic proposition underpinning law is the question of how and why law can give its subjects an explanation for obeying it (Finnis 2020). Ever since Aristotle, the expected response to this question was human nature's striving for the good, for personal happiness, or for a collective "common good." We obey law because we recognize that a good legal system enables us to flourish as human beings. One could see this as the precursor of biological behaviorism's assumption that it is in our nature to make rational choices. Many legal theorists embrace this position, including the most well-known legal positivist H.L.A. Hart, whose so-called "minimal" version of natural law suggests that we need law to protect ourselves and overcome our fundamental vulnerability and finitude, i.e. that through law we ensure access to water, food, and shelter. A less "minimal" relation between human nature and law is formulated by Christian thinkers, including Thomas Aquinas, who perceives law as emerging from our natural, God instilled, capability for reasoning and logical deductions (for Aquinas Reason is what enables us to discover the laws of God's Creation). As a consequence, if human made laws do not follow God's divine creation, they are un-natural and cannot be just: bad law necessarily pertains to evil, unlawful conduct and, hence, does not need to be obeyed ("an unjust law is not law"). Hart and Aquinas occupy the two ends of the natural law spectrum; a secular, positivist position seeing law as a survival aid and a religious, all-encompassing position seeing law as reflecting Divine Creation. While Hart's law aims to organize society effectively through legitimate procedures, Aquinas' law aims to overcome our original sin while guiding humans in how to best fulfill God's plan.

The history of natural law shows how two distinctive ordering practices—nature and law—merge into forms of knowledge that enable political, economic, and legal justifications to constrain, free, punish, regulate, and organize the human and natural world. In the Western collective imagination, our "humanity" produces the moral and cultural order of civilization, replacing the chaotic, wild, and dangerous *natural* state of nature. Hence, the *nature* of "natural law" functions as an instrumental truth justifying what law can or cannot do; law represents a particular form of metaphysical knowledge about (our) nature, it is this knowledge which then *naturalizes* the (cultural) legal order.

The role of nature in the social contract theories of the seventeenth and eighteenth centuries led to a more concrete political and legal formulation of natural rights, most prominently declarations of constitutive rights as exemplified in the French and American Revolutions. What this history shows is the *liminal* character of the law/nature dynamic. It is the law that symbolizes and solidifies the line between nature and culture, between anarchy and order, instincts and civilization, between good and evil; yet the law is also constituted in relation to nature. The legal scholar David Delaney writes:

> Just as nature remains both threat and foundation, so the essentially liminal quality of law is perpetuated. Nature and law always limit each other: law limits – confines, constrains, controls – nature, but never completely. There is always a residue and so always the real possibility of regression, of falling back, of losing control. Nature marks the limits of the power and reach of law (even if the location of the boundary changes, its function as boundary remains). No matter how strongly law is identified with the human side of the line, with nomos as against physis, nature and law are always in contact as it were. Nature never goes away and is never far from the center of concern. (see p. 93 in 2003)

In short, law needs nature as its opposite to establish its own authority over and toward the human; law determines and defines what "nature" can or cannot be.

As such the "nature of law" and the "laws of nature" both draw from a particular idea of nature and law at a particular time and place. This "particular" is informed by metaphysical and normative claims (about what nature or law really is or prescribes us to be). It is David Hume's "empirical skepticism" that trims back these claims to what can be experienced, observed, extracted, or deduced from experimental testing and observation. This new empirical attitude replaces the prior "natural" universal order. Immanuel Kant quickly undermines the idea of objective observations by showing that our knowledge of the world is a synthesis of our experiences with our reasoning and imaginative capacities. The dispute between "realism" and "idealism," between the perception of nature as independent and external or as intrinsic to the operations of our mind, echoes similar debates in law: Is law intrinsic to all humans or an arbitrary social construction? In short, from Hobbes to Locke, and Kant to Hegel, philosophical speculations about human nature translate into different political, moral, and legal ideologies, practices, and social science "programs," determining how law and political institutions best reflect our "real" nature (to fight each other, to reason, be good, be free—and so on). There is a constitutive circularity emerging here: ideas about human nature produce practices (of how humans ought to act) while these practices reenforce existing ideas (of what humans' nature amounts to).

We now focus on several "turns" in the relation between law and biology that coincide with the shifts in natural law discussed earlier: the political, proprietary, realist, and economic turn. Each "turn" reflects a genealogical shift in how nature and law relate, mutually impact each others' "order," and how this further affects our self-understanding and ways we organize society.

The Political Turn: The Nature of Sovereign Authority

Thomas Hobbes' *Leviathan* (1996 [1651]) marks a crucial shift from classical natural law to modern political philosophy and to universal rights. Hobbes also leaves behind a natural philosophy of human existence (*Naturphilosophie*) to pursue a more *political* philosophy of human community. The core of Hobbes' theory is the idea that *fear* governs all our actions and decisions. Avoiding situations of fear, like war or death, presents the most fundamental motivation for humans to create law and obey sovereign power and governance. For Hobbes the Sovereign should *not* be controlled by constitutional limitations because the fear sustained by a certain level of arbitrariness acts as strategic instrument to enforce obedience. Hobbes understanding of nature also implies that citizens are not bound by any natural moral duty, but rather, possess a natural *right* for self-preservation, freedom, and pursuit of self-interest. This moves nature from being the prime source of morality to being the prime motivation for legal rights.

Replacing moral duty with legal rights, Hobbes reformulates the idea that we are determined by (and ought to obey) the "laws of nature" with the idea that we are determined by politics (and ought to obey) the "laws of civil government." In other words, Hobbes replaces the "laws and morality of nature" with the "nature of politics and law" (whereby "nature" now refers to the internal principles, rational, and purpose of politics and law). This leads to the intertwining of *secular*, *political*, and *juridical* notions of Sovereign power. Hobbes' well-known statement that "the law of nature and the civil law contain each other and are of equal extent" (see p. 138 in Hobbes 1996 [1651])[1] affirms that the nature of law in the civil sphere is a politics of nature, which constitutes the nature of politics.

In sum, Hobbes *Leviathan* replaces classical natural law with a process of *political legitimation*. This, together with the growing influence of an *empirical* worldview is—at least pre-Darwin—the most crucial intervention in nature's relationship to law. One could argue that these shifts anticipate and condition the later development of sociobiology and evolutionary models of jurisprudence. Indeed, today's dominant "Law and Economy" movement with its idea that all our actions are driven by freedom or fear is deeply rooted in Hobbesian thinking translated into the language of behavioral economy or contracts (freedom) and adjudicated through liabilities or tort law (fear). The freedom/fear logic drives and rationalizes all human behavior and our obedience or disobedience to the law. The contemporary countermovement of Law and Political Economy (LPE) aims to re-politicize the relation between law and economy showing how "rising inequality and precarity, political alienation, the entrenchment of racial hierarchies and intersectional exploitation, and ecological and social catastrophe"[2] result from the capitalist economization of law and deep anchoring of power and financial structures in law.

The Proprietary Turn: A Natural Right to Own

While John Locke mostly affirms Hobbes' political conversion of natural law into absolute rights, he offers a different understanding of the scope of those rights. For Hobbes, the right for self-preservation is unlimited, turning the "state of nature" into a "state of war" that calls forth a social contract mediated by a sovereign authority to contain the conflict. Locke, on the other hand, concludes that natural rights also evoke natural *duties* (like the need to respect the rights of others):

> The *State of Nature* has a Law of Nature to govern it, which obliges everyone: and Reason, which is that Law, teaches all Mankind, who will but consult it, that being all equal and independent, no one ought to harm another in his Life, Health, Liberty, or Possessions. (Locke 1960: 271)[3]

Since our capacity to reason should lead to one categorical conclusion (don't do unto others what you don't want done unto you), conflict should naturally be avoided. Hence, if law aligns with our natural capacity to reason, and all reasoning obliges us to obey the "Law of Nature," conflicts can be avoided as long as our duty toward each other aligns with the nature of our reasoning. Once Locke establishes a secular connection between natural rights and natural duties it allows a new connection between rights and property:

> Though the earth, and all inferior creatures, be common to all men, yet every man has a property in his own person: this nobody has any right to but himself. The labor of his body, and the work of his hands, we may say, are properly his. Whatsoever then he removes out of the state that nature hath provided, and left it in, he hath mixed his labor with, and joined to it something that is his own, and thereby makes it his property. It being by him removed from the common state nature hath placed it in,

> it hath by this labor something annexed to it, that excludes the common right of other men. For this labor being the unquestionable property of the laborer, no man but he can have a right to what that is once joined to, at least where there is enough, and as good, left in common for others. (Locke 1960, 287)

Although ascriptions of work and labor to persons or states go back many centuries, and were, in various forms, formalized since Roman law, the idea that the "labor" of my body and hands are "properly" mine is Locke's radical intervention in the relation between law and nature. More crucial than the direct connection between body, labor, and property is the claim that if we remove something "from the common state of nature" and labor is "annexed to it," the "common right of other men" must be excluded, and "no man but he can have a right to what that is once joined" (see p. 287 in Locke 1960). Locke does add an important caveat to individualized property rights: "at least where there is enough, and as good, left in common for others." A simple example of the contentious impact of Locke's property concept is described eloquently by Susan Silbey in her *Invocations of Law on Snowy Streets* (2011). Here she points to an everyday practice of ownership on the snowy streets of Boston where in winter people clear the street in front of their house from snow, put a chair on the shoveled parking spot, and claim it as *theirs*. There is no law that legalizes this claim, but it is enacted as an unspoken rule due to the internalization of Locke's law of property in our collective legal consciousness.

A darker aspect of Locke's property theory is revealed in his own connection to the slave trade. Not only was Locke secretary of the Lords Proprietors of the Carolinas, a British colony still allowing slavery, he also owned stock in slave trading. But Locke's position is ambivalent. In his seminal *Two Treatises of Government* Locke distinguishes between legitimate and illegitimate slavery; the only legitimate form of slavery was mastery over captives after winning a just war. Illegitimate slavery is obedience created and enforced by Absolute Monarchy because it deprives all people, including Afro-American slaves, from freedom, property, and rights. Hence, although Locke was indirectly connected to the practice of slavery, his political ideas seemed to clearly oppose enslavement outside the context of a justly won war.

Locke's original theory of property forms the hinge between the state of nature (the human condition before state governance installs its rule over humans), and civic government. As outlined above, for Locke (see p. 287 in Locke 1960 [1698]) "every Man has a *Property* in his own *Person*" whereby the "Labour of his Body, and the *Work* of his Hands" are "properly his." This also applies to what we can take from nature; although God provides us with nature for Locke, once man "mixed his *Labour* with" nature, even by simply picking a berry from a tree, nature becomes "something that is his own" thereby excluding "the common right of other Men."[4]

This notion of property still reverberates in a complex manner in contemporary intellectual property rights, among others, in the biosciences. Is a genetically manipulated mouse, or a reproduced cell line from a human, still nature or, by manipulation or reproduction, turned into a person's or laboratory's property? And how does the individually manipulated mouse differ from the genetic code of the mouse distributed as pure data among scientist or used to invent an entirely new mouse?[5] While Locke did not invent the concept of intellectual property, his justificatory logic still influences how property claims, intellectual ownership, or patent rights are made. As we will describe in more detail in Chapter 4, Locke's concept of property also determines the biolegal claims made for one's "own" DNA or notion of personhood, and in Chapter 6 for those pertaining to the notion of parenthood. As political thinkers both Hobbes and Locke radically reformulate how "human nature," emerging from a "state of nature," must find ways to live together peacefully by creating a social contract and state authority which must reflect our natural right to survive, be free, equal, safe, and own the product of our labor.

The Realist Turn: Law as the Nature of Experience

Between the natural-legal nexus and the social-legal nexus emerges the prolific concept of the "body-politic" (*corpus politicum*). Originating in classical Greek philosophy (Plato), and central to both Roman and Byzantine state formation, the concept informs the idea of incorporation (first used by the Catholic church, further developed in Islamic thought, and culminating in Hobbes' *Leviathan*) and symbolizes the synthesis of the state and its people. The genealogy of the "body politic" is more complex but for our argument here it suffices to point out, as Thomas Lemke does, how the metaphor of the "body" in the nineteenth century conflates political interventions with medical interventions:

> The organicist concept understands the state not as a legal construction whose unity and coherence is the result of individuals' acts of free will but as an original form of life, which precedes individuals and collectives and provides the institutional foundation for their activities. The basic assumption is that all social, political, and legal bonds rest on a living whole, which embodies the genuine and the eternal, the healthy, and the valuable. The reference to "life" serves here both as a mythic starting point and as a normative guideline. Furthermore, it eludes every rational foundation or democratic decision-making. From this perspective, only a politics that orients itself toward biological laws and takes them as a guideline can count as legitimate and commensurate with reality. (Lemke 2011: 10)

The merging of the laws of nature and the laws of a "common life," now perceived as one giant collective organism, are not just used for a metaphorical "political body" but, rather, to constitute a "politics of biological life" as the actual task of the "body politic"—of the all-encompassing state authority.

Throughout the eighteenth and nineteenth centuries, moral and political philosophy continued debating the universal nature (or "natural universality") of moral principles, liberty, and rights. But by the end of the nineteenth century, most Western jurisprudence moved away from natural law to embrace different forms of "soft" or "hard" legal positivism (Bentham, Austin, Hart), realism (Holmes), or pluralism (Posner). We have no space here to examine all these perspectives but instead focus more closely on Olivier W. Holmes' legal realism and the "turn" away from the foundational grounding of legal substance or reasoning in natural law toward a more experiential and pragmatic understanding of law. Instead of political ideas about law as governance, we should observe and examine law as a concrete practice. As a child of his times, Holmes embraces Darwin's evolutionary model of environmental adaptation and the "survival of the fittest" as a method which can also explain the effectiveness and persistence of certain laws, where others lose their meaning and disappear.

In current jurisprudence, Oliver Holmes' legal realism represents the transition from the natural-legal to the socio-legal view, whereby the social is increasingly explained in evolutionary terms. For Holmes law does not reflect nature as such, rather, it reflects the natural structure of how society changes and evolves. Replacing "nature" with "realism" shifts law from moral questions and principles (impartiality, equality, justice) to scientific and positivistic questions and "facts." This "realism," in turn, introduces new forms of observing and understanding what law *does*, thereby incorporating economic, behavioral, and anthropological viewpoints into law.

For Holmes, law is in constant flux, adapting to the circumstances of life and society. While in substance legal realism dismisses any natural foundation of law, in method it embraces the examination of nature as conceptualized by biology and its main theoretical explanatory framework: evolution through adaptation. In other words, legal realists attribute to law a sort of natural pragmatism. The "bad-man" is not interested in principles of justice or guilt, but only in what he can "get away with." Law's main usefulness is not moral condemnation or correction but how best to recognize, regulate, and predict human behavior; due to his fear of legal punishment the "bad man" should not commit the crime he plans to commit, otherwise law is ineffective. This reasoning, of course, brings "nature" back into the equation: to know when punishment functions as a sufficient deterrence is to assume we know something about human nature.

Holmes' experiential model of law's evolution merges with economic theories which follow the same logic: if law—just like "nature" and the "market"—is analogous to the "law" of capitalist competition and supply/demand curves, then legality partakes in the same "rationality" as natural and economic "necessities." In this narrative, new laws or precedents simply satisfy a demand. As the next section explains,

Frederic Hayek and the very influential Chicago School of "economy and law"—which, besides Hayek, contained well-known economists and lawyers like Milton Friedman, George Stiglitz, Robert Vogel, and Richard Posner—were exactly located at this nexus of economic, legal, and biological reason. Any form of the greater good, natural inclination for cooperation, or social empathy is essentialized to utilitarian behavior. Thus, it is Holmes' pragmatic-evolutionary "realism" which prepares the grounds for the merging of law, sociobiology, and economics, affirming the crucial bifurcation between natural law and legal positivism:

> [I]t is finally for science to determine…the relative worth of our different social ends…it is our estimate of the proportion between these, now often blind and unconscious, that leads us to insist upon and to enlarge the sphere of one principle and to allow another gradually to dwindle into atrophy. (see p. 462 in Holmes 1899)

While Holmes rejects natural law as a set of universal values which the law should reflect, he embraces the scientific notion of natural evolutionary progress wherein adaptation determines species survival. In this way laws are not formulated—and cases are not decided—on higher principles or even constitutional measures. Instead, law is naturalized as a historical, developmental process determined by the "felt necessities" of the time. These "necessities" indicate that law only changes—unfolds—in response to certain social needs, while other needs slowly fade away. One can pose the question of what is experienced as "necessary" is not simply determined by what has most power or authority at any given time. Thus, Holmes replaces the foundational force of nature with a normative contingent position that, ironically, justifies itself by adopting a scientific theory of nature as key explanatory framework for changes in law; law, just like nature, must adapt to new circumstances to survive. This realist-pragmatic viewpoint of law as basically demand-driven inspires the new field of "law and economy" merging a quasi-neutral, and quasi-scientific, process of "natural" self-regulation in which "free" markets and "free" laws meander together into the future (whereby this evolutionary/needs-driven idea of law and economy aims to depoliticize the power of state authority while finance and political elites solidify their privileged positions).

The Economic Turn: The Nature of Human Behavior

At the end of the nineteenth century, a more problematic use of evolutionary theory informs the new field of criminology. Through its focus on the study of "moral insanity," criminology relied on the biological reasoning that humans were driven by instinctive nature. In line with the emerging discourse of Social Darwinism, criminality was given an evolutionary scant through the belief that the criminal's "moral sense must be in a state of decay, causing them to devolve or go backward down the evolutionary ladder (...) [i]t seemed clear that the problem lay in offenders' brains; moral insanity was a biological condition, not a social problem" (see p. 19 in Rafter 2009). By the end of the nineteenth century, criminologists influenced by scientific positivism embrace the power of numbers and the creation of quantitative models. Moreover, developments in psychiatry and medicine influence the formative works of the German psychiatrist Richard von Krafft-Ebing and the Italian psychiatrist Cesare Lombroso, often considered the founding fathers of criminology. Throughout the nineteenth century, criminology used biology as a tool to profile criminal characteristics, mostly through the now discredited science of phrenology, which links the suspect's visual appearance to criminal minds and nature.

Poignantly, this biologization of law led to a hegemonic economic perspective on law: nature as biology becomes a means to predict human behavior in apparently rational and calculable terms. The foremost representative of these ideas is the American behavioral psychologist B.F. Skinner, who popularized experimental observations of human behavior under controlled circumstances. Working against psychoanalytical insights—and according to critics thereby dehumanizing psychology—Skinner's behavioralism contends that external stimuli trigger particular responses, and human behavior is conditioned by its consequences (this recalls Hume's theory of moral sentiments: we repeat what causes pleasure and avoid what causes pain). If crime is not punished (causing pain) it will increase (due to being rewarded); also, if we see others getting away and profiting from crime those observing this are triggered to act similar. Hence, Skinner's work promises a better diagnosis of why people commit crimes, when authority is respected or rejected, and what policies, regulations, and enforcements are best to control individuals and group behavior.

The field of criminology emerges in the context of industrial capitalism and the liberal state, both of which understand the individual as agential and self-sufficient (offering choice while demanding self-reliance). This was perceived as both promise and threat. Both capitalism and the liberal state promote themselves in egalitarian terms. These "rational" natural scientific methods provide an explanatory justification for inequality at the level of biological type for the individual, and group and evolutionary process at the level of social order. Law then serves as a site of application and mediation between state and capital where these methods are deployed to manage populations without implicating the social structure itself.[6]

The economic turn in law became increasingly captivated by rational choice theories produced within behavioral sociology and cognitive psychology. Deploying quantitative and experimental methods derived from the natural sciences, these disciplines portrayed themselves, and were accepted by the social and political elite, as methodical, objective, and precise. The rise of sociobiology and cognitive psychology since the 1970s has further strengthened the idea that law and criminology can be based on a scientific perception of human behavior as rational, predictable, and calculable. These ideas spread at the turn of the twentieth century, informing much of the eugenic thinking which we discuss in the next section.

Critical Times: The Legacies of Social Darwinism

Accompanying the birth of criminology, and its reliance on evolutionary theories, are scientific theories of race and of degeneration. Criminology propagated the idea that some humans evolve, while others degenerate toward (moral) insanity, explaining why they resort to crime or savagery. While early criminologists believed in the reversibility of degeneration, arguing that criminals could improve their chances through education and proper living conditions, the end of the nineteenth century witnessed a hardening of hereditarianism (Meloni 2016; Rafter 2009). No longer was criminality seen as a social problem that was resolvable by social address; instead, criminality was now "fixed in degenerates' germ plasm," merging with the broader discourse of what became known as scientific racism (see p. xix in Rafter 2009). As criminologist Nicole Rafter shows:

> Scientific racism often became part of the explanation of criminal behavior: the worst offenders are like savages, closer to Hottentots and other black-skinned primitives than to normal white people, with their well-evolved moral sense. In this case as in that of degeneration theory, explanations based on ideas about evolution pointed to a biological problem: criminals' brains must be more primitive than those of well-evolved whites. (ibid.)

Ideas of scientific racism—gathered under the general banner of Social Darwinism—took root in individualist societies. In Europe these ideas merged with and contributed to the nationalist and imperialist ideologies germinating into fascist movements. Hence, the early twentieth century witnessed the emergence of eugenic thinking instrumental to German Nazism. The violent and genocidal consequences of the latter were quintessentially illustrated by the Holocaust but had roots in colonial occupations throughout the world. Academically, the new fields of biological anthropology and ethnology played a key role in establishing "Völkerkunde" (population/ethnic studies) in Germany and the US aiding the legitimization of eugenic thinking more generally.

Eugenics and Scientific Racism

As historian Gretchen Schafft (2002) illustrates, Nazi anthropologists were profoundly caught up in delivering scientific "evidence" of racial differences and traits. Working predominantly for the Kaiser Wilhelm

Institute (KWI), their studies on genetics and human heredity informed the eugenic practices of the Third Reich, mostly through the racial "classification" of people. Their scholarship and teaching (to SS doctors) about racial hygiene reinforced Nazi eugenic laws and policies. Many anthropologists continued this work after the end of the war. According to Schafft, this was possible because anthropologists in other places too—and also before the war—were involved in producing the same kind of scholarship and conceptual tools (see pp. 130–131 in 2002). Put directly, eugenic thinking was not unique to fascism or Nazism.

As early as the beginning of the nineteenth century, forms of negative eugenics were present in many parts of Western society, represented foremost by forced sterilization practices and laws (Meloni 2018: 71). Fears of racial degeneracy informed sterilization laws in the US and Europe, first in Switzerland, and swiftly followed by Scandinavian countries and Germany (ibid.). It was only in the second half of the twentieth century that such sterilization programs were discontinued. According to the political sociologist Maurizio Meloni, managing reproduction is at the hinge joining negative and positive eugenics: "given the often-irrational form taken by reproduction in human societies, the process had to be controlled and rationalized so to avoid undoing the work of natural selection" (Meloni 2018: 72).

Sociobiological Explanations

Although eugenic research programs were discontinued after the Holocaust and human rights thinking inserted in national laws and international treatises, the persistence of evolutionary theories in law since the 1970s was exemplified by the influence of the new subdiscipline of sociobiology in Europe (Meloni 2018), North America (Schwartz 1982), and the Antipodes (Ardill 2009). Introduced by the natural scientist Edward Wilson, sociobiology was another iteration in the long trajectory of evolutionary theories, where the indisputable nature of humans (and their place in society) can be observed and explained by scientific discovery. According to the legal scholar Donald Elliott, the sociobiological school of legal evolution sees evolution not merely as a metaphor for the internal dynamics of a legal system (like Holmes' legal realism assumes); its members believe that evolution is the *causal* process that accounts for the existence of law and, to some extent, for the law's form and content. What distinguishes sociobiological theories of legal evolution is not the claim that law evolves (in evolutionary fashion), but the claim that law is itself a product of evolution (is evolved naturally from nature). Because sociobiology seeks to establish biological explanations for human and animal behavior using deductive methods, it can be regarded as a scientific research program that treats and analyzes law as bioscientific fact (Rafter 1997).

Feminist thinking was also implicated in sociobiological theories. Since the 1970s, feminist struggle focused not only on male power and dominance in the social, economic, and political realm, but also challenged patriarchal claims of men's "stronger" biology (referring to physical and mental justifications for male social and domestic dominance). The question of biological sex fragmented the feminist movement (a split still reverberating in current debates about transsexuality): if biological sex is the basis of "the female experience" this essential difference must recognize a female worldview and reformulate our social, political, and legal relations through an essentialist lens beyond the question of human equality (MacKinnon 1989). But if biological difference is instead a social construction affirmed through the performance of conventional masculine and female gender roles (Butler 2002), this construction is just a dominant script that must be rewritten and adjusted to our actual sense of self and identity. This triggers the problem of whether law's task is to protect and empower the essentially different experiences of women (Habermas 1995), or instead, should recognize and offer protective rights to all gender identities (or both). The ongoing public debates about sex and gender—biology or social construction— destabilize social and legal identity statuses. Biolegality, as a heuristic lens, aids in clarifying how these entanglements unfold through an analysis of bio(techno)logy's impact of sex and gender (Oudshoorn 2003; Roberts 2002).

An important distinction must be made between the "Law and Biology" movement and the development and dissemination of sociobiology. The subfield of "Law and Biology," as Donald Elliot describes, can better be understood as the "Law and Ethology" movement, which argues that "law is a particular type of animal behavior, and therefore that some of the same scientific tools that have been developed to study other forms of animal behavior can be adapted to studying the animal behavior that we call law" (see pp. 596–597 in 1997). Against determinist ideas, or thoughts of Social Darwinism, this line of thought emphasizes the centrality of empirical work around animal behavior to be useful for thinking about the internal structures of law (see p. 597 in Elliot 1997).

Primatologist Frans de Waal extends this line of thinking to show that "human" forms of politics, morality, empathy, and law can be traced back to group dynamics in primates, hence, revealing a *continuum* instead of a false *binary* between "wild" animals and "civilized" human behavior. For De Waal humans are highly ambivalent creatures because:

> Being both more systematically brutal than chimps and more empathic than bonobos, we are by far the most bipolar ape. Our societies are never completely peaceful, never completely competitive, never ruled by sheer selfishness, and never perfectly moral. (see p. 212 in De Waal 2005)

The study of celebrity animals like primates to better understand our own behavior should not hide the fact that most animals on the planet are legal property while being caged and slaughtered for human consumption; our own "social Darwinism" clearly overrides the academic attribution of similarity or intelligence to our co-species. Between explaining criminal or consumption behavior sociobiology now influences both law and economy experts, offering in equal measure knowledge on how humans cause harm or shop.

In his book *The Human Measure*, Donald Kelley (see p. 279 in 1990) aptly describes the ongoing tension within the social and human sciences between a qualitative and quantitative approach. These tensions can also be observed in law and many point to how law deploys the "conceptual temptations of biologism, evolutionism, racism, psychologism, functionalism, structuralism" to declare itself a science (ibid.). In the next chapter (Chapter 4) we further explore this tension as we move further into the twentieth century and the age of genetics. Genetics allowed biology to become even more firmly entrenched on positivist and quantitative scientific terrain. Its impact on law was decisive and requires another qualitative analysis of biolegality. While this chapter focused on the longer historical narrative between law and nature and law and biology, the biotechnological revolution precipitated by the emergence of genetics is the focus of the next chapter.

Conclusion

The internal paradox of natural law, as our short genealogy shows, is that law both legitimizes itself as *not*-nature (not violent chaos) while simultaneously justifying legal-political authority *from* nature (it is our nature to be free, fearful, self-interested, own our labor, and so on). This chapter's genealogical overview of "law and biology" is far from complete, but hopefully offers a lens through which the now often invisible traces of "nature" and "biology" in our contemporary legal concepts and practices are highlighted to reveal a continuity of the entanglement between law and biology.

Notes

1. More precisely, this quote can be found in chapter 26, paragraph 4.
2. See https://lpeproject.org/, accessed September 6, 2023. Thanks to Ben Stork for providing this reference.
3. John Locke, Two Treatises of Government, Ed. Peter Laslett (Cambridge: Cambridge UP, 1960), §6, p. 271. *Second Treatise of Government*.
4. Ibid., Second Treatise, §27, p. 288.
5. I thank Ben Stork for this apt observation.
6. Again, many thanks to Ben Stork for pointing to and describing this historical context, which we have copiously included here with permission.

References

Ardill, Allan. 2009. Sociobiology, Racism and Australian Colonisation. *Griffith Law Review,* 18(1), pp. 82–113. https://doi.org/10.1080/10383441.2009.10854631.

Braverman, I., ed. 2017. *Gene Editing, Law, and the Environment: Life Beyond the Human.* London: Routledge.

Butler, Judith. 2002. *Gender Trouble.* London and New York: Routledge.

Delaney, D. 2003. *Law and Nature.* Cambridge: Cambridge University Press.

De Waal, Frans. 2005. *Our Inner Ape.* New York: Riverhead Books.

Elliott, D.E. 1997. Law and biology: the new synthesis. *Saint Louis University Law Journal, 41*(2), pp. 595–624.

Finnis, John. 2020. Natural Law Theories. In Zalta, E. N. (ed.), *The Stanford Encyclopedia of Philosophy* (Summer 2020 ed.). https://plato.stanford.edu/archives/sum2020/entries/natural-law-theories/.

Gieryn, T. 1983. Boundary Work and the Demarcation of Science from Non-Science: Strains and Interests in Professional Ideologies of Scientists. *American Sociological Review,* 48(6), pp. 781–795.

Habermas, J. 1995. Paradigms of Law. *Cardozo Law Review* 17, p. 771.

Hobbes, T. 1996 [1651]. *Leviathan* (ed., R. Tuck). Cambridge: Cambridge University Press.

Holmes, O. W. 1899. Law in Science and Science in Law. *Harvard Law Review,* 12(7), pp. 443–463.

Huffman, Carl. 2021. *Alcmaeon: The Stanford Encyclopedia of Philosophy* (ed., N. Zalta). https://plato.stanford.edu/archives/sum2021/entries/alcmaeon/.

Jasanoff, S. 1997. *Science at the Bar: Law, Science, and Technology in America,* vol. 9. Cambridge: Harvard University Press.

Jasanoff, S. 2001. Ordering Life: Law and the Normalization of Biotechnology. *Politeia,* 62, pp. 34–50.

Kelley, D. R. 1990. *The Human Measure: Social Thought in the Western Legal Tradition.* Cambridge: Harvard University Press.

Lemke, T., 2011. *Biopolitics. In Biopolitics.* New York University Press.

Locke, John. 1960 [1698]. *Two Treatises on Government.* Cambridge: Cambridge University Press.

Lynch, M., Cole, S. A., McNally, R., and Jordan, K. 2010. *Truth Machine: The Contentious History of DNA Fingerprinting.* Chicago: University of Chicago Press.

MacKinnon, Catherine A. 1989. Sexuality, Pornography, and Method: Pleasure Under Patriarchy. *Ethics*, 99(2), pp. 314–346.

Meloni, M. 2016. *Political Biology: Science and Social Values in Human Heredity from Eugenics to Epigenetics*, p. 295. Basingstoke: Palgrave Macmillan.

Meloni, M. 2018. A Postgenomic Body: Histories, Genealogy, Politics. *Body & Society*, 24(3), pp. 3–38.

Oudshoorn, Nelly. 2003. *Beyond the Natural Body: An Archaeology of Sex Hormones*. London and New York: Routledge.

Pottage, A. 2007. The Socio-legal Implications of the New Biotechnologies. *Annual Review of Law and Social Science*, 3, pp. 321–344.

Rafter, N. H. 1997. *Creating Born Criminals*. Urbana: University of Illinois Press.

Rafter, N.H. ed., 2009. *The origins of criminology: A reader*. London: Routledge

Roberts, Celia. 2002. A Matter of Embodied Fact Sex Hormones and the History of Bodies. *Feminist Theory*, 3(1), pp. 7–26.

Schafft, Gretchen E. 2002. Scientific Racism in Service of the Reich: German Anthropologists in the Nazi Era. *Annihilating Difference: The Anthropology of Genocide*, pp. 117–136. Berkeley: University of California Press.

Schwartz, Richard D. 1982. On the Prospects of Using Sociobiology in Shaping the Law: A Cautionary Note. *Journal of Social and Biological Structures*, 5(4), pp. 325–333.

Silbey, S. S. 2011. J. Locke, op. cit.: Invocations of Law on Snowy Streets. *Journal of Comparative Law*, 5(2), pp. 66–91.

3

The Biotechnology Revolution

Introduction

While human societies have long employed nature technologically to assist them in agriculture, food production, and medicine[1], it is the late twentieth century's development of genomic science and bioinformatics that gives biotechnology its particular currency today. The emergence of genomic science revolutionized biotechnology, leading to the myriad applications and understandings currently in circulation, while the discovery of gene splicing in the 1970s—cutting DNA from one organism and inserting a gene from another—paved the way for genetic engineering and recombinant DNA. These technologies were used in fields as disparate as marine and aquatic culture (blue biotechnology), agriculture and environment (green biotechnology), medicine including reproductive medicine (red biotechnology), industrial biotechnology (white biotechnology), even insinuating themselves into governance and law through practices like forensics (DNA fingerprinting). Indeed, while biotechnology might conventionally refer to any technological application of biological systems or living organisms to make or modify products or processes—the United Nations' definition[2]—it, in fact, encompasses much more; conceptually, biotechnology must also

S. van Wichelen and M. de Leeuw, *Biolegality*, Biolegalities,
https://doi.org/10.1007/978-981-99-8749-8_2

include both biological knowledge and its applications, the techniques that make such applications possible, and the politics, economics, and legalities that allow them to flourish (see also Pottage 2007).

This chapter probes the dilemmas that biotechnology brings to law through the prism of evidence and regulation. First, we evaluate how genetics and other biotechnical methods play a pivotal role in our legal institutions, pointing to how the introduction of DNA changed methods of determination—the processes through which evidence for a particular identification problem is gathered—in a wide range of legal fields. We emphasize how revolutionary techniques in DNA "fingerprinting"—the scientific technique that uses DNA to determine the identity of a person—granted new authority to scientific evidence in the courtroom, especially in criminal cases. Similarly, and more recently, contemporary neuroscience—and its application through neurolaw—is emerging as a new and consequential form of lawful evidence based on scientific authority. However, in both instances, critics argue that establishing the validity of these new forms of scientific evidence—what concepts, contexts, methods, and knowledge regimes they utilize—requires social science research. And, as scholarship from anthropology, sociology, history, philosophy, human geography, and science and technology studies (STS) shows, the status of scientific evidence is both methodologically or procedurally *unstable*. It is also conditioned by a certain politics and legal culture, along with economic forces.

While genetic technologies maintain and extend existing criminological practices of determination, other new bioscientific paradigms and applications radically challenge our conceptions of *biology* and *law itself*, reconceptualizing living matter and troubling the very foundation of legal forms. One could even say that society and law first had nature to reconcile with, then biology, and now biotechnology. With the biotechnological turn society is creating new relations between nature and biology and seeing how the social (law) and biological (life) are entangled (see p. 255 in Pottage 2004 and p. 269 in Hamilton 2012). These discussions within and outside of law reveal that foundational legal forms are becoming frail, and cracks are forming that demand attention not only from legal scholars, but from society as a whole.

These cracks are triggered by a range of scientific insights and technologies (gene editing, synthetic biology, microbiome research, epigenetics, etc.) that, despite their diversity and range, all intervene in the very notion of biological life. What is called the New Biology, or the "postgenomic biosciences," disturbs legal concepts based conventional understandings of nature and biology. These postgenomic developments revise the legal premises of personhood (De Leeuw and Van Wichelen 2020; see also Van Beers 2017; Pottage 2002, 2007); who and what law defines as "persons" becomes uncertain and fragmented, necessitating a refashioning of the legal categories of "persons" and "things" (see p. 3 in Pottage and Mundy 2004).

Hence, the second part of this chapter turns to biotechnology's bearing on regulation by describing the shift in bioscience from *knowing* life (for instance through the decoding of the human genome) to *making life* (e.g., by creating new biological forms through synthetic biology). With this shift, our capacities to tinker with life take on momentous proportions impacting the way we see and understand ourselves as human and more-than-human as well as our relationship to nature and non-nature. Thus, at the heart of genetic engineering there is a tension within the nature of the natural. For instance, in agriculture the genetic modification of crops puts into question whether the resulting tomato or corn can still be defined as "natural." With these and other uncertainties, the governance of life emerges as a central predicament in law and society, which we explore through cases in the US, Europe, and beyond the Global North. As we will show, capital and markets, along with states and sovereignties, are pivotal to discussing and illustrating the stakes of biotechnology in contemporary global society.

Tools of Determination

As described in our previous chapter, the combination of behavioral biology and cognitive psychology was central to law in late twentieth century, where they functioned as the legitimate sciences of culpability and intentionality in criminal law. However, the advent of genetic testing brought on significant changes to the tools of determination used in criminal law and, by extension, immigration, and family law. As genetic technologies foundationally altered methods of inference, their uses in law also demonstrated the endurance of legal knowledge. Biotechnology, then, does not *necessarily* change legal forms: the "apparatus of accusation" may have changed (Pavlich 2016) but the legal form remains intact. As such, the analogy of evolutionary adaptation still strongly informs the social dimensions of legal systems for legal realists and beyond. In this sense, we agree with Pottage and Mundy, that "law is the original biotechnology", meaning that law is the original authority of defining nature (see p. 39 in Pottage and Mundy 2004). While the scientific ability to splice genes in the 1970s, and the subsequent uses of DNA in courts, significantly changed the legal apparatus, the interpretation and application of evidence by legal experts maintained the dominant status quo.

DNA fingerprinting now replaces the more arduous, and less reliable, standard technique for fingerprinting with what Jasanoff describes as, "fingerprinting on steroids" (see p. xx in Jasanoff 2010). In a relatively short time, courts embraced DNA profiling, making it one of the most important instruments in forensics where it is considered "a virtually failsafe tool, if properly used, for linking violent crimes to the persons who perpetrated them" (see p. xx in ibid.). The widespread use of DNA evidence in courts, having improved considerably since its inception, exists in constant tension with earlier forms of evidence after leading to the closure of many criminal cases as well as the overturning of wrongful convictions. Presented as an ultimate form of proof, DNA evidence superseded conventional forms of evidence such as testimony and suspect confessions. As Lynch et al. contend (2010), DNA evidence is broadly understood by the lay public, especially when combined with its slick, high-tech representation in popular forensic procedural television, as a "truth machine," conveying the message that forensic biotechnologies are exact and truthful.[3]

More recently, neuroscience is finding a stronghold in criminal law too. With the rapid growth of neuroscience, as we describe more fully in Chapter 5, novel conceptions in the field of neurolaw emerge through the representative and symbolic force that the brain holds in society, namely as center of control and agency in the individual. Despite its sophistication and complexity, the impact of neuroscience on legal practice, in fact, owes much to a commonsense logic that obscures "the considerable sociotechnical work necessary to align the practices and concerns of neuroscientists with those of lawyers" (see p. 30 in Pickersgill 2011). This commonsense logic rests on the belief that biology delivers certainty about behavior and can, therefore, authoritatively assist in prosecuting or exonerating criminal suspects. The scientific expertise that lawyers and judges call on when judging causal claims about a suspect's genes or brains are remarkably similar to nineteenth-century practices of collecting evidence: the term DNA fingerprinting itself echoes the practice of fingerprinting pioneered by Francis Galton in the 1860s, while brain scanning echoes the physiognomic photography of Cesare Lombroso.[4] Understanding DNA testing and brain scans within this longer history of evidence collection, shows the way probabilistic bioscience takes on a far more deterministic meaning within the framework of the courtroom (Jasanoff 2001; Rose 2010). In this sense, DNA testing and brain scans extend the criminological paradigm; even as they radically change methods of determination, these technologies do not necessarily change legal form, setting them apart from developments in other arenas of biotechnology.

Futures of Life-Making

Following the Holocaust there was a deep rejection of biology as selective destiny, and more broadly, as the basis for the nation-state and national identity. Rather than eugenics, biology became redefined by a biomedical discourse that increasingly determined how citizens articulated their rights and duties, enacting a "politics of life itself" characterized by risk management, the molecularization of life, and by somatic and corporeal individuality in excess of state institutions (Rose 2006). Society moved away from the state in its relationship with biology and toward the market as the site of anticipatory citizenship.

In the second half of the twentieth century biotechnology more and more took the form of a promissory science, one full of anticipatory discourse (Pickersgill 2011; Stevens 2016). As controversies around biotechnology emerge (for instance around reproductive technologies, GMOs, cloning, etc.) tensions are articulated in terms of conflicts and risks, conjuring "socio-technical imaginaries" that understand societies as informing the design of how technology ought to aid the future of the nation-state (Jasanoff and Kim 2015). At the heart of managing conflicts and risks—especially their entanglements with markets and finance profiting from risk—is regulation: law and legality stands in a recursive relationship to the experimental infrastructures of biotechnology, constantly adjusting how to order its "potentially disordering" affects (Foucault 1977). As such biolegality, as a conceptual approach, emphasizes the expressively hybrid nature of regulation. Rather than advancing a form of reductionism via regulation, the encounter between biotechnology and law results in a proliferation of identities and rights.

Some anthropologists argue that the changes in biotechnology and the biosciences are changing the concept of nature or biology itself (Rabinow 1996; Strathern 2005; Franklin 1995). While groups of people can be characterized in both social and biological ways, the biological and the social are in fact mutually constituting. Genetic testing in medicine, for instance, introduces new kinds of ethical dilemmas for conditions that have not, and may never surface (like increased risk for breast cancer or Alzheimers). If genetics as social paradigm is fueling the rise of biosociality (Hacking 2006; Rabinow 1999), it is also creating—as we discuss further in Chapter 6—new forms of biological citizenship where biological conditions culminate in rights claims and new modes of civic engagement (Rose and Novas 2005; Petryna 2004).

Law plays a crucial part in negotiating these novel definitions of life, and subsequently of death, itself. For example, the medico-legal parameters demarcating the assessment of life and death vary significantly across jurisdictions, with many countries adopting brain death as legal death; thereby natural death is superseded by a social and legal fiction, which facilitates potential post-death medical practices such as organ donation. Similar debates can be observed about the boundaries of birth via "wrongful life" claims. In contrast to "wrongful birth" claims, where a mother sues a medical professional for failing to prevent pregnancy, "wrongful life" claims are made by children with disabilities, who appeal to a right not to be born. Some states in the US and Europe allow disabled persons born via assisted reproduction to sue medical institutions for wrongful life.[5] As the legal scholar Van Beers describes, these cases create new legal fictions: "Children can legally contest the natural state in which they were born, this creates a legal persona that precedes their conception. The disjunction between legal and biological temporality is brought to new heights" (see p. 586 in Van Beers 2017). These legal configurations of life and death illustrate law's attempt to engage, negotiate, and incorporate medical technologies within law's own practical understanding of life. In doing so, longstanding social and legal conventions for defining the boundaries of life are given new meaning by the application of biotechnological advances.

These rearrangements of law and life are not confined to (human) medicine. On the contrary, biotechnology's impact on agriculture, food, and the environment is immense and the globalization of these technologies has restructured societies around the world, fundamentally altering the ways communities care for their land, grow crops, and produce food. The technological changes are facilitated (if not forced) by structural adjustment programs tied to development loans from the World Bank and the International Monetary Fund, organizing production toward the global market rather than local need. When we look at agricultural regulations across countries and regions, we see different delineations of biology and nature. In the US, a plant product modified with gene editing, but without interference or inclusion of foreign DNA, is not categorized as a Genetically Modified Organism (GMO). European regulations, which stress the technical process, or the *how* of modification, on the other hand, stipulate that any interference with the original plant through biotechnology is classified as a GMO regardless of the source of the DNA involved. The debate around GMOs becomes part of the ways political cultures self-consciously forge nation-states and national identities (Jasanoff 2005; Parthasarathy 2011). This forging of different political cultures works in part because Europe (as a supranational entity) is a relative equal to the US, but as some of our cases throughout Part II will illustrate, the functions and meanings of biology in society (and for the idea of nation-state) work differently relative to poorer/weaker countries in the Global South or rival countries such as China.

In *The Future of Human Nature* (2003) the social philosopher Jürgen Habermas argues that biotechnology radically problematizes the structure of legal form by collapsing the categorical distinction between the *made* and the *grown.* Defining human nature as that which is grown, namely given and contingent, and "the made" as that which humans choose to do and control, Habermas' deliberations of this problem in law articulates humanist concerns to preserve the grown against the made in order to uphold the species' capacity for autonomous self-realization. A collapse of the made and the grown in the legal domain would compromise the integrity of humanness by technical manipulation, the latter which seems to be detached from the domain of nature.

But as we explained in Chapter 2, this pitting of biotechnology against nature (and law to control nature) goes against the now well-established and conclusive proposition that both science and law are products of culture and the culture of nature (Rabinow 1992; Jasanoff 2001; Delaney 2003). Moreover, opinions differ on *how* biotechnological knowledge alters legal knowledge, depending on the context and understanding of law. Scholars associated with "new kinship studies," for instance, claim that the introduction of DNA testing challenges the very foundations of family law (Franklin 2013; Dolgin 1997; Strathern 2005). However, analyzing how genetics reconfigure legal motherhood, fatherhood, and parenthood reveals how new forms do not just appear; they are dynamically negotiated using authority and legitimacy from both science and law. Law, then, is not simply overruled by science; rather, an active reworking takes place between legal and scientific knowledge (Jasanoff 2001; Strathern 2005). Biotechnology does not alter the instrumentalist understanding of law as a means to an end. Instead, we could follow Pottage's critique of Habermas, and argue that new scientific truths merely "[*loosen*] the ontological consistency of the grown [and *reveal*] the sense in which the grown was produced and stabilized by legal norms and institutions" (see p. 324 in Pottage 2007, emphasis added). It is this *loosening* and *revealing* that is at the heart of biolegality's operation.

A good example of this *loosening* and *revealing* can be found in scrutinizing the status of embryos within the context of ownership or property, which changes depending on their materiality and context, thereby destabilizing the legal distinction between persons and things. According to anthropologist Marilyn Strathern the potentiality of (legal) personhood for an embryo varies according to whether the embryo is frozen or thawed, implanted in utero or discarded (see p. 175 in Strathern 1999). Strathern notes that the question of whether body parts can be seen as property is dependent not only on biological relations (the detachment of parts from the whole), but also on conceptual relations (the conceptual capacities to detach parts from wholes). While biotechnology—such as reproductive technologies—enables the detachment of body parts from their wholes, law can qualify the conceptual process of detachment, which continues the oscillation between persons and things.

Such *loosening* and *revealing* becomes especially poignant as we enter the "postgenomic" era, which emphasizes instability rather than stability and gene-environment interactions rather than genetic reductionism (Richardson and Stevens 2015; Lappé and Landecker 2015). Jenny Reardon (2017) addresses the postgenomic *condition* in terms of the historical and discursive break initiated by the completion of the human genome project and its failure to produce the conclusive causal relations between genes and diseases the project promised. The postgenomic era, then, signals a move away from the more deterministic set-up of the old genetics and looks to find more hopeful outcomes in a new biology. Drawing from the social study of postgenomic biosciences (Helmreich 2015; Mansfield and Guthman 2015; Meloni 2018; Reardon 2017; Rajan and Leonelli 2013; Richardson and Stevens 2015; Stallins et al. 2018; Van Wichelen and Keaney 2022), we investigate the implications of postgenomic practices and visionary projections for law and legal theory.

Postgenomics marks a fundamental shift in analysis and knowledge production by initiating a new phase of doing, organizing, funding, and structuring research (Richardson and Stevens 2015). The main driver of the postgenomic condition is the massive expansion and intensification of sequencing technologies and capacities, introducing multiple forms of uncertainty in the status of biomaterials and bioinformation. These technologies lead to new challenges in determining the valuation, classification, and provenance of materials and data, and to their providers, annotators, and curators (Ankeny and Leonelli 2015). While sequencing technologies convert materials to data, the analysis of this data becomes part of an aggregate shared across (private and public) practitioners, which becomes increasingly undistinguishable in terms of locating data's origins or trajectories, raising legal and economic concerns around ownership. At the same time, the computing power of metagenomic[6] data systems enable the commodification and disclosure of unique information about individuals, raising privacy concerns similar to and potentially greater than those raised by human genetics information collected from blood and swabs.

The emergence of what has been termed the *-omics* revolution is representative of this technical turn in the postgenomic era. Microbiomics, epigenomics, proteomics, metabolomics, transcriptomics, and even reproductomics are all examples of large-scale computational research projects based on sequenced biological samples. These new sequencing technologies—associated with shotgun sequencing and the development of metagenomics—radically changed the biosciences, transforming the legibility, value, and calculability of knowledge (Leonelli 2015; O'Malley and Skilling 2018). As bioinformatics becomes the primary activity (over collecting and culturing) in bioscience, ethico-legal challenges of scientific exchange move from materials to information.

For many social theorists, postgenomics is philosophically characterized by the promise to go beyond the genome (Meloni 2016; Keller 2015). Epigenomics, for instance, emphasizes how the biological is inextricably entangled with the social and the environmental. While the study and application of epigenomic sciences remain susceptible to deterministic views of disease (Mansfield 2012; Mansfield and Guthman 2015), the discovery of epigenetic plasticity also informs new biosocial models. What kind of friction occurs when biological plasticity is used as a form of legal justification? With respect to microbiomics, the human microbiome is, it turns out, a more-than-human or multi-species affair, involving the life and labor of tiny organisms—trillions of microbes—inside the human body, in a dynamic relationship with environmental conditions outside of the body (Helmreich 2015; Lorimer 2019; Paxson 2008). What is redefined here is not the subject herself, so much as the more-than-human assemblage constituting the microbiome, including the natural and built environment (dirt, chemicals, pollution), diet (food and nutrition), medicine (anti-microbials), and organisms both inside (microbiota or hookworms) and outside bodies (pets and wildlife). For the study of microbiomics, the more-than-human microbiota inside and outside our bodies inform the status not only of human wellbeing, but also of ecological wellbeing since plant, soil, and aquatic microbiomes interact with human microbiomes. Hence, these biosocial understandings change the very idea of human individuality, leaving uncertain the legal status of persons or the "things" that make up the "person."

Roberto Esposito's ideas around the fundamental distinction between persons and things in modern Western law are especially relevant to posthuman legal theory (2015). Esposito aims to upset this constitutive distinction at the point where law provides the foundational basis of what he terms "immunity"—as the antithesis of community and develops a normative, affirmative biopolitical thesis that emphasizes bodies that individually or collectively resist closure (see p. 25 in Brown 2019). In effect, destabilizing the distinction between persons and things becomes central to realizing an affirmative biopolitics that "break[s] the prohibitions that have always blocked re-unifying life and lives in their singular and impersonal dimensions" (see p. 114 in Campbell and Luisetti 2010). As Hanafin explains, this view allows a particular conception of the body, one that is permeable and that resists the "dispositif of the person and the biopolitical governance of lives through law" (see p. 18 in 2014). Drawing on socio-legal studies and affirmative biopolitics (Amendola 2012; Hanafin 2013, 2014; Vatter and De Leeuw 2023), we will in the next chapters of Part II theorize how a possible "third person" or a notion of "bodyhood" could further disturb the modern Western legal distinction between persons and things and address some of the emerging postgenomic quandaries.

The Sway of Economy

Perhaps the most crucial qualitative changes biotechnology brought to society over the past decades are what Rabinow and Rose refer to as "new circuits of bioeconomics." They write:

> [A] large scale capitalization of bioscience and mobilization of its elements [are brought] into new exchange relations: the new molecular knowledges of life and health are being mapped out, developed and exploited by a range of commercial enterprises, sometimes in alliance with States, sometimes autonomous from them, establishing constitutive links between life, truth and value. (see p. 203 in Rabinow and Rose 2006)

Contemporary scholars of biocapital, biovalue, and bioeconomies (Cooper 2011; Rajan 2006, Pavone and Goven 2017; Waldby and Mitchell 2006; Rose 2007; Cooper and Waldby 2014) have identified a new terrain that neoliberalism is expanding into, constitutive to the developments of bioscience and biomedicine. The emergence and stimulation of global bioeconomies, in which life is increasingly predicated on risk assessments, speculation, and experimentation, opens the body and its parts to the vagaries of financial capital, commodification, and forms of biological labor. A leading explanation for the fundamental change in kind—and not only degree—is the immense impact of the privatization of science (and the legal tools that come with securing this process) on the organization and regulation of biotechnology in contemporary society (Drahos and Braithwaite 2001; Mirowski 2011; Rajan 2006). For instance, the emergence of a bioeconomy was significantly enabled by a legal decision in the US allowing the patenting of genetically modified microorganisms, seen by many as the first legal move to commercialize what was seen as belonging to nature.[7] This recent history of legal governance is central to the investigation of biolegality.

The historical conditions of advanced or late capitalism and the development of technologies that reconfigure the political subject have led some posthuman theorists of capital to contest the centrality of governance in the enactment of biopower or the reliance on anthropocentric notions of the human body as the focus of biopolitics and governmentality. In *Life as Surplus*, Melinda Cooper (2011), for instance, demonstrates how uncertainty and risk transforms the regulation of bodies in ways that depart from Foucault's description of the political administration of populations and the uses of demographic methods as governmental techniques. While the central impetus for a Foucauldian biopolitics was the paradigmatic transformation from disciplinary society to governmentality, the new form of advanced capitalism described by Cooper is only loosely connected to the realm of intentional governmental policies, and instead, is attached more firmly to financial calculations of risk and insecurity. In showing how the neoliberal economy of financialized capitalism in the twenty-first century informs a new political economy, Cooper does not dismiss the heuristic power of biopolitics in examining these changes but encourages us to become much more attentive to new speculative modes of growing the bioeconomy. This development challenges the causal determinations in the biopolitical model of governmentality (the changes in the epistemological and discursive constitution of populations and their management that drive historical change). In Cooper's revision, biopolitical ordering is itself fundamentally shaped and driven by the dynamics of global capital.[8]

There is an operational and metaphorical fusion between the growth of biological life (tissue, cells, genes, microbes) and the growth of economic life, which marks contemporary capitalism as a different regime of biopolitics from the Fordist era. Moreover, by re-applying a Marxist model to this new regime of biopolitics, Cooper points to the corollary effect of scarcity in promoting a capitalism that embraces an excess of regenerative life. Following the new post-industrial economy, biological life transforms into surplus value, but not without devaluing life; a constitutive element of the neoliberal regeneration of life is the structural devastation of life. Apart from real life consequences, this devastation also destabilizes the idea of the human as a separate and ontologically different type of life. Capital in the age of neoliberalism and biotechnology enmeshes with biological life in ways that reconfigure or go beyond life as *zoe* (biological or bare life) and *bios* (qualified citizen life which can form a biography). As life is inserted into the neoliberal logics of capital, it mutates into populations, statistics, casualties, collateral damage, data, or surplus. A biolegal approach utilizes this perspective on biocapital to push forward the argument that law—as much as science or markets—is constitutive of the proliferation of bioeconomies.

This bioeconomic development extends to a global scale. The economization of nature and biology are globalized through the expansion of intellectual property rights via the 1995 introduction of the "trade-related aspects of intellectual property rights" (TRIPs) agreement. Advocated by American economists and lawyers, TRIPs commercialized science on a global scale through the consolidation of intellectual property (IP) instruments. The expansion of IP rights from the North (though particularly from the US) to other parts of the world ultimately failed to incentivize infrastructural development that could improve self-sufficiency in health care delivery, bioscience innovation, and localized drug discovery in developing nations (Drahos and Braithwaite 2001; Sell 2003).

In sum, as a technoscience, biotechnology is deeply entangled in the contours of economic theory (Callon 1998) and legal scholars—alongside economists—are expected to reassess the forms by which they define and regulate these developments. This means examining biotechnology as part of the knowledge economy, where information becomes increasingly decoupled from its material, biological source (Franklin 2001). It also means observing how vitality and biovalue are (legally) engineered in the laboratory and how this corresponds to value generated in the broader economy. As we increasingly move toward the world of bioinformatics, focus needs to be directed to how the exchange of information generates value and how this exchange is enabled by our regulatory frameworks. As this book proceeds with describing such instances, it is important to recognize that the new biotechnologies are instruments that, in coproduction with law, regenerate nature and biology; this biolegal condition, constitutive to defining nature itself (Rabinow 1999; Thacker 2006), is responsible for the rise of a new set of relations between biotechnology and society that produce novel ideas of "nature," "biology," "identity," and "rights."

Conclusion

This chapter engaged with the recent history of biotechnology starting with the rise of molecular genetics prompted by Watson and Crick's discovery of the structure of DNA. These developments had tremendous impacts on society and law leading to a profound transformation in criminology and forensic sciences. We examined whether and how biotechnology is driving a paradigmatic shift in law. We demonstrated that law is responding to these changes in two ways: applications of the biosciences and biotechnology that extend existing legal practices and applications that are radically challenging our legal concepts and forms.

Notes

1. The term is largely believed to have been coined in 1919 by Hungarian engineer Károly Ereky.
2. UN Convention on Biological Diversity, Art. 2.
3. Investigations of court cases in the US and the UK by Lynch et al., however, reveal that DNA evidence is much more controversial than portrayed in popular media and culture (see also Edmond 2002). Many cases raise ethical and legal questions about the ways in which biotechnology and bioinformation are being embedded in forensic processes. The occurrence of human errors, insufficient DNA material, and the problematic uses of probabilities all attest to the instabilities of DNA evidence. Lynch et al. describe how statistics and probabilities are vital components in the integration of genetic technologies in forensic practice. And in the field of DNA testing, the calculations that produced probabilities were often skewed. Forensic experts consulted in courts give a probability (say one in 3 billion) that the DNA match was an error, but Lynch et al. demonstrate how such probabilities are highly dependent on the context, for instance, in terms of the boundaries of a "reference population" (see also pp. 23–25 in Prainsack 2010). These contexts are often indiscernible in testimonies of experts.
4. Galton was not only the founder of the fingerprinting technique; the cousin of Charles Darwin, he was also the first to study and suggest that "mug shots" were important in deciphering types of criminals. Inspired by Galton's work, Lombroso developed the field of positivist criminology, in which physiognomic photography played a central role in identifying criminal behaviors with physical features and defects.

5. In a wrongful life case, the child with disabilities seeks damages for the suffering in life because of their creation. While such cases have not been very successful, they have piqued considerable interest within legal philosophy as the cases address the temporality of personhood and the morality of procreation.
6. In contrast to genomic analysis, which examines the complete genetic information of one organism, metagenomic analysis examines DNA from multiple organisms. This often takes place directly from environmental or clinical samples and are difficult to accomplish without sequencing methods.
7. On June 16, 1980, the US Supreme Court ruled in the case of Diamond v. Chakrabarty that a bacterium, capable of breaking down crude oil, could be patented by the Indian-born Ananda Chakrabarty who worked for General Electric. As Chakrabarty's work involved entire organelles—rather than direct gene manipulation—this established a new precedent, which was seen by many as commodifying nature.
8. Many thanks to Ben Stork for this acute addition to the discussion.

References

Amendola, A. 2012. The Law of the Living: Material for Hypothesizing the Biojuridical. *Law, Culture and the Humanities*, 8(1), pp. 102–118.

Ankeny, R. and Leonelli, S. 2015. Valuing data in postgenomic biology. In Richardson, S.S. and Stevens, H. eds. *Postgenomics: Perspectives on biology after the genome*. pp. 126–49, Durham: Duke University Press.

Brown, Wendy. 2019. *In the Ruins of Neoliberalism: The Rise of Antidemocratic Politics in the West*. New York: Columbia University Press.

Callon, Michel. 1998. Introduction: The Embeddedness of Economic Markets in Economics. *The Sociological Review*, 46(1), pp. 1–57.

Campbell, T. and Luisetti, F. 2010. On Contemporary French and Italian Political Philosophy: An Interview with Roberto Esposito. *Minnesota Review*, 75(1), pp. 109–118.

Cooper, M. E. 2011. *Life as Surplus: Biotechnology and Capitalism in the Neoliberal Era.* Washington: University of Washington Press.
Cooper, M. and Waldby, C. 2014. *Clinical labor: Tissue donors and research subjects in the global bioeconomy*. Duke University Press.
Delaney, D. 2003. *Law and Nature*. Cambridge: Cambridge University Press.
De Leeuw, M. and Van Wichelen, S. 2020. *Brave New Law: Personhood in the Age of Biolegality*. Cham: Palgrave Macmillan.
Dolgin, J. L. 1997. *Defining the Family: Law, Technology, and Reproduction in an Uneasy Age*. New York: NYU Press.
Drahos, Peter and Braithwaite, John 2001. Intellectual Property, Corporate Strategy, Globalisation: TRIPS in Context. *Wisconsin International Law Journal*, 20, p. 451.
Edmond, Gary (2002) "Legal Engineering: Contested Representations of Law, Science (and Non-Science) and Society," *Social Studies of Science* 32(3): 371–412.
Esposito, R. 2015. *Persons and Things: From the Body's Point of View.* New York: Wiley.
Foucault, M. 1977. *Discipline and Punish: The Birth of the Prison* (trans. A Sheridan). London: Allen Lane.
Franklin, S. 1995. Science as Culture, Cultures of Science. *Annual Review of Anthropology*, 24(1), pp. 163–184.
Franklin, S. 2001. Biologization Revisited: Kinship Theory in the Context of the New Biologies. In Franklin, S. and McKinnon, S. (eds.), *Relative Values: Reconfiguring Kinship Studies*, pp. 302–325. Durham: Duke University Press.
Franklin, S. 2013. *Biological Relatives: IVF, Stem Cells and the Future of Kinship.* Durham: Duke University Press.
Fukuyama, F. 2003. *Our Posthuman Future: Consequences of the Biotechnology Revolution.* New York: Farrar, Straus and Giroux.
Habermas, J. 2003. *The Future of Human Nature*. Cambridge: Polity Press.
Hacking, I. 2006. Genetics, Biosocial Groups & The Future of Identity. *Daedalus*, 135(4), pp. 81–95.
Hamilton, Jennifer. 2012. The Case of the Genetic Ancestor. In Wailoo, K., Nelson, A., and Lee, C. (eds.), *Genetics and the Unsettled Past: The Collision of DNA, Race, and History*. New Brunswick: Rutgers University Press.
Hanafin, P. 2013. Rights, Bioconstitutionalism and the Politics of Reproductive Citizenship in Italy. *Citizenship Studies*, 17(8), pp. 942–955.
Hanafin, P. 2014. Resistant Lives: Law, Life, Singularity. *Soft Power*, 1(1), pp. 91–107.

Helmreich, S. 2015. *Sounding the Limits of Life: Essays in the Anthropology of Biology and Beyond*. Princeton: Princeton University Press.

Jasanoff, S. 2001. Ordering Life: Law and the Normalization of Biotechnology. *Politeia*, 62, pp. 34–50.

Jasanoff, S. 2005. *Designs on Nature: Science and Democracy in Europe and the United States*. Princeton: Princeton University Press.

Jasanoff, S. 2010. Foreword. In Hindmarsh, R. and Prainsack, B. (eds.), *Genetic Suspects: Global Governance of Forensic DNA Profiling and Databasing*. Cambridge: Cambridge University Press.

Jasanoff, S. and Kim, S. H. 2015. *Dreamscapes of Modernity: Sociotechnical Imaginaries and the Fabrication of Power*. Chicago: University of Chicago Press.

Keller, E. F. 2015. The Postgenomic Genome. In Sarah S. and Stevens, H. (eds.), *Postgenomics: Perspectives on Biology After the Genome*, pp. 9–31. Durham: Duke University Press.

Lappé, M. and Landecker, H. 2015. How the genome got a life span. *New genetics and society*, 34(2), pp. 152–176.

Leonelli, Sabina. 2015. What Counts as Scientific Data? A Relational Framework. *Philosophy of Science*, 82(5), pp. 810–821.

Lorimer, J. 2019. Hookworms Make us Human: The Microbiome, Eco-Immunology, and a Probiotic Turn in Western Health Care. *Medical Anthropology Quarterly*, 33(1), pp. 60–79.

Lynch, M., Cole, S. A., McNally, R., and Jordan, K. 2010. *Truth Machine: The Contentious History of DNA Fingerprinting*. Chicago: University of Chicago Press.

Mansfield, B. 2012. Race and the New Epigenetic Biopolitics of Environmental Health. *BioSocieties*, 7, pp. 352–372.

Mansfield, B. and Guthman, J. 2015. Epigenetic Life: Biological Plasticity, Abnormality, and New Configurations of Race and Reproduction. *Cultural Geographies*, 22(1), pp. 3–20.

Meloni, M. 2016. *Political Biology: Science and Social Values in Human Heredity from Eugenics to Epigenetics*, p. 295. Basingstoke: Palgrave Macmillan.

Meloni, M. 2018. A Postgenomic Body: Histories, Genealogy, Politics. *Body & Society*, 24(3), pp. 3–38.

Mirowski, P. 2011. *Science-mart: Privatizing American Science*. Cambridge: Harvard University Press.

O'Malley, M. A. and Skillings, D. J. 2018. Methodological strategies in microbiome research and their explanatory implications. *Perspectives on Science*, 26(2), pp. 239–265.

Parthasarathy, S. 2011. Whose Knowledge? What Values? The Comparative Politics of Patenting Life Forms in the United States and Europe. *Policy Sciences*, 44, pp. 267–288.

Pavlich, G. 2016. Avowal and criminal accusation. *Law and Critique*, 27(2), pp. 229–245.

Pavone, V. and Goven, J., eds. 2017. *Bioeconomies*. Cham: Palgrave Macmillan.

Paxson, H. 2008. Post-pasteurian cultures: The microbiopolitics of raw-milk cheese in the United States. *Cultural anthropology*, 23(1), pp. 15–47.

Petryna, A. 2004. Biological Citizenship: The Science and Politics of Chernobyl-exposed Populations. *Osiris*, 19, pp. 250–265.

Pickersgill, M. 2011. Connecting Neuroscience and Law: Anticipatory Discourse and the Role of Sociotechnical Imaginaries. *New Genetics and Society*, 30(1), pp. 27–40.

Pottage, A. 2002. Unitas personae: on legal and biological self-narration. *Law & Literature*, 14(2), pp.275–308.

Pottage, A. 2004. Introduction: The Fabrication of Persons and Things. In Pottage, A. (ed.), *Law, Anthropology, and the Constitution of the Social: Making Persons and Things*. Cambridge: Cambridge University Press.

Pottage, A. 2007. The Socio-legal Implications of the New Biotechnologies. *Annual Review of Law and Social Science*, 3, pp. 321–344.

Pottage, A. Mundy, M. 2004. eds. Law, Anthropology, and the Constitution of the Social: Making Persons and Things. Cambridge Studies in Law and Society. Cambridge University Press.

Prainsack, B. 2010. Key issues in DNA profiling and databasing: Implications for governance. In Hindmarsh, R. and Prainsack, B. eds., 2010. *Genetic suspects: Global governance of forensic DNA profiling and databasing*. pp.153–174, Cambridge: Cambridge University Press.

Rabinow, P. 1992. Artificiality and the Enlightenment: From Sociobiology to Biosociality. In Crary, Jonathan and Kwinter, Sanford (eds.), *Incorporations*. New York: Zone Books.

Rabinow, P. 1996. Artificiality and Enlightenment: From Sociobiology to Biosociality. In *Essays on the Anthropology of Reason*, pp. 91–111. Princeton, NJ: Princeton University Press.

Rabinow, P. 2010. Artificiality and Enlightenment: From Sociobiology to Biosociality. *Politix*, 2, pp. 21–46.

Rabinow, P. and Rose, N. 2006. Biopower today. *BioSocieties*, 1(2), pp. 195–217.

Rajan, K. S. 2006. *Biocapital: The Constitution of Postgenomic Life*. Durham: Duke University Press.

Rajan, Kaushik Sunder and Leonelli, Sabina. 2013. Introduction: Biomedical trans-actions, Postgenomics, and Knowledge/Value. *Public Culture*, 25(3), pp. 463–475.

Reardon, Jenny. 2017. *The Postgenomic Condition: Ethics, Justice, and Knowledge After the Genome.* Chicago, IL: University of Chicago Press.

Richardson, Sarah S. and Stevens, Hallam, eds. 2015. *Postgenomics: Perspectives on Biology After the Genome*. Durham: Duke University Press.

Rose, N. 2006. *The Politics of Life Itself: Biomedicine, Power, and Subjectivity in the Twenty-First Century*. Princeton: Princeton University Press.

Rose, N. 2007. Molecular Biopolitics, Somatic Ethics and the Spirit of Biocapital. *Social Theory & Health*, 5(1), pp. 3–29.

Rose, N., 2010. 'Screen and intervene': governing risky brains. *History of the human sciences*, 23(1), pp.79-105.

Rose, N. and Novas, C. 2005. Biological Citizenship. *Global Assemblages: Technology, Politics, and Ethics as Anthropological Problems*, pp. 439–463. Oxford: Blackwell Publishing.

Sell, S. 2003. *Private Power, Public Law: The Globalization of Intellectual Property Rights*. Cambridge: Cambridge University Press.

Stallins, J. Anthony, Law, Derek M., Strosberg, Sophia A., and Rossi, J. J. 2018. Geography and Postgenomics: How Space and Place are the New DNA. *GeoJournal*, 83(1), pp. 153–168.

Stevens, H. 2016. *Biotechnology and Society: An Introduction*. Chicago: University of Chicago Press.

Strathern, M. 1999. *Property, Substance, and Effect: Anthropological Essays on Persons and Things*. London: Athlone Press.

Strathern, M. 2005. *Kinship, Law and the Unexpected: Relatives Are Always a Surprise*. Cambridge: Cambridge University Press.

Thacker, E. 2006. *The global genome: Biotechnology, politics, and culture*. MIT press.

Van Beers, B. 2017. The Changing Nature of Law's Natural Person: The Impact of Emerging Technologies on the Legal Concept of the Person. *German Law Journal, 18*(3), 559–594.

Van Wichelen, S. and Keaney, J. 2022. The Reproductive Bodies of Postgenomics. *Science, Technology, & Human Values*, 47(6), pp. 1111–1130.

Vatter, Miguel and De Leeuw, Marc. 2023. Human Rights, Legal Personhood and the Impersonality of Embodied Life. *Law, Culture and the Humanities*, 19(1), pp. 106–125.

Waldby, C. and Mitchell, R. 2006. *Tissue economies: Blood, organs, and cell lines in late capitalism.* Duke University Press.

Part II

Biolegal Rearrangements

4

Genes: Exchange and Property

> Corporealization involves institutions, narrative, legal structures, power-differentiated human labor, technical practice, analytic apparatus, and much more. The processes 'inside' bodies — such as the cascades of action that constitute an organism or that constitute the play of genes and other entities that go to make up the cell — are interactions, not frozen things…a word like gene specifies a multifaceted set of interactions among people and nonhumans in historically contingent, practical, knowledge-making work.
>
> Donna Haraway, *Modest_Witness@Second_Millenium.*

Introduction

The last few decades have witnessed a dash to propertization of (human and non-human) genes by biotech companies and research consortia seeking to capitalize on these materials for further research and profit. At the same time, knowledge about genes and genetics plays an increasingly prominent role in society. Ever since the human genome project, the power of genetics—and later genomics—have had a tremendous

S. van Wichelen and M. de Leeuw, *Biolegality*, Biolegalities,
https://doi.org/10.1007/978-981-99-8749-8_3

impact on how we think, feel, and relate to ourselves and to one another. Genetics call upon and alter our desires and imaginaries, whether in terms of our kin, our selves, or our communities (Nelkin and Lindee 1995). With authoritative genetic metaphors, and its movement to synecdoches (from the figurative to the literal), come new forms of legal claims and rights.[1] Proprietary assertions have come from individuals, who seek control, via ownership or use, of their genetic material and information, and from groups or nation-states, whose cultural or sovereign claims include the protection of genetic resources.

While studies on patents, property, and ownership may describe different legal contexts and jurisprudence, we deliberately bring these debates and legal conceptualizations together in order to reconcile a larger anthropological problem, namely that of exchange in contemporary societies undergoing rapid advances in biotechnology and law. This chapter draws predominantly on the literature demonstrating how gene patents or ownership of genetics are experienced, appropriated, or resisted, and how genetic property produces new normative ideas about what can be termed "biological exchange." By studying forms of exchange—including conceptualizations around gifts, commodities, and more recent concepts such as "benefit-sharing"—the chapter describes how the biolegality of genetic materials produces novel anthropological concerns.

The emergence of what has been called "the new genetics," or postgenomics, poses distinct complications for the question of ownership and exchange. The prefix "new" here denotes the aftermath of the human genome project, when knowledge of genes opened up the potential to predict disease, leading to new diagnostics and treatments, and, conceptually, to the medicalization of health conditions through the science of genetics (Finkler 2000). Since the late 1990s, revolutionary technologies in genomic sequencing have accelerated and the acrruing knowledge has had significant consequences for how people relate to genetic information.

This chapter describes the stakes of the new genetics in different biolegal terrains. First, on the level of economy and value, genes are seen as providing science with vital information about the human body. As new innovations using genetics proliferate, genetic material has become increasingly valuable. Bioscience and biomedicine are contested arenas for disputes over whether genetic material and information should be publicly available for research or patented by companies and institutions for the benefit of Research and Development (R&D) units. These units have increasingly become important sites in what is often called the global bioeconomy. Characteristic for the twentieth and twenty-first centuries is the conviction that science and the utilitarian uses of bioscientific enterprises (pharmaceuticals being one of the important sites) cannot evolve without the protection of patents and other forms of intellectual property such as copyright and trademarks (Sunder Rajan 2006, 2017). At stake then is the value of invention on the one hand, and the influence of economic interests on the other. While context-bound—playing out differently for instance in European or US contexts—the politics of patents exemplify a biolegal story involving the value of genes and the meaning of nature.[2]

Second, genes have a distinct connection to the idea of self and ownership vis-à-vis their "own person" as formulated by John Locke's central legal dogma that "every man has a property in his own person" (1698). Contemporary cases in the Euro-American contexts highlight the persistent weight this doctrine holds in the public domain, even as it is increasingly up for debate in the legal domain. This debate revolves around the growing fragmentation of personhood precipitated by the so-called *molecularization* of life and bodies (Rose 2001; Rabinow and Rose 2006; Raman and Tutton 2010; Niewöhner 2011)—the compartmentalization and extraction of genetic material and information—which produce complex situations in the legal arena. Analyzing the informal case of "Henrietta Lacks" and the formal legal case of *Moore v. Regents of the University of California* case, we demonstrate the complexities of ownership claims on biological material and information. In doing so, we discuss how the stakes of biolegal claims are irreducible to the subject of privacy, but instead also illustrate the importance of bodily integrity for social and political functions beyond the individualized notion of privacy. Central to the debate of legal personhood is the idea of property and ownership as embedded in modern law's moral-legal distinction between persons and things. Carefully unpacking this modern division, we look at how new biosciences trouble the distinction between persons and things by reconfiguring not only the "human," but also "nature" and "the natural," opening new questions around the commodity and commodification that resonate with older questions in anthropological theory.[3]

Advancements in genomic sequencing, bioinformatics, and synthetic biology further complicate these fragmentations by engaging non-natural and modified materials or extracted information (including genomic sequence data). The third section of this chapter engages with bioinformation and the proliferation of genetic sequence data in the bioscientific world and beyond. Genomic sequencing has replaced the exchange of genetic material with genetic data, prompting companies, universities, and states to think about the nature of scientific and commercial exchange as tangible materials become intangible.[4] This shift complicates debates around property and ownership since the ontology of biological material is now scrutinized. Often, the value of bioinformation becomes clear only after it has been used in research or commercialized by industry. This creates problems when providers of genetic resources call for adequate acknowledgment or compensation.

How to democratically deliberate and regulate the new conditions of bioinformation and exchange in accord with conceptions of social justice on a global and planetary scale, is the final question of this chapter. In thinking about scientific exchange transnationally, issues of biosovereignty have become more prominent in the past few decades and legal frameworks that control biological resources therefore become effective political instruments. The sovereign power of nation-states to capitalize on and/or to protect genetic resources has informed debates about global inequality since the late 1980s, when science and industry made vast investments in collecting genetic specimens for potential medical and commercial use—also known as bioprospecting—especially from the Global South. Here we discuss how biolegal arrangements inherent to benefit-sharing instruments—those measures ordained by international law to secure economic benefits to those countries or communities whose intellectual or biological resources are being exploited—have emerged alongside these debates and are currently being reconfigured by the datafication of biological (genetic) material and the fungibility of bioinformation. Anthropological reflections on these biolegal developments in ownership and genomics highlight the social and political intricacies of exchange in contemporary global society.

Patent Politics and the Biolegality of Invention

Examining the history of patents, the legal historian of science Mario Biagioli (2006, 2011) describes how—around the 1790s—a shift occurred that transformed a patent system based on privileges to one based on rights. This shift came as inventors were increasingly expected to include a so-called "specification of the invention." This specification came at a time when absolutist regimes were being replaced by representative ones, which necessitated a different justification of the patent system; by making the invention known to the public, the monopoly of the patent becomes, so to speak, "harmless" and therefore justified within a context where people were starting to gain political representation. But these specifications produced more than mere descriptions of inventions:

> They created the conditions of possibility for the development of an entire legal, administrative, and political discourse and related institutional practices about patents and inventions. Patent specification, in fact, was not just as a concept related to others (as in an episteme, a paradigm, or a conceptual scheme), but a concept inseparable from a specific practice—a kind of inscription that both fostered and needed to be managed by a bureaucracy like that of the patent office… The very existence of patent law as a separate branch of legal doctrine was made possible by the textualization of invention—by the coming to be of an inscription about which a discourse could be developed, debated, copied, and transmitted. (Biagioli 2006: 1160)

What Biagioli was able to show in this remarkable reconstructive work is that a seemingly mundane legal-administrative step (specification of invention) was not only able to transform a system of privilege into one more publicly accountable; it also changed the logic of patents:

> The practice of disclosure—turning patents into knowledge and then making that knowledge public so that it could eventually flow into the public domain—was part of the process that made inventors into citizens: that is, individuals with rights. Intellectual property rights and right-bearing citizens were constructed at the same time, through the same process. (ibid.)

It is helpful to understand current patent politics from this more archaeological framework as it puts into perspective the conditions in which radical shifts may emerge. These are not necessarily found in the domain of politics or policy; but through operations of legal materialities and their constitutive interaction with society.

The more recent debate about whether or not human genes can be patented is a case par excellence to study how today's patent politics over biotechnological "objects" are never about one "thing," namely an object that we understand as stable or immutable (see Carolan 2010). Rather, in the biotech age, it is the mutability of biological material—core to its ontological identity—that in its relationship to intellectual property rights forms new conditions for "conflict, inequality and injustice" (see p. 126 in Carolan 2010). Echoing biolegal disputes that started with the patenting of plants in the early twentieth century (Sherman 2008) and culminating in patenting battles over Monsanto's global monopoly on seeds at the turn of the century (Carolan 2010; Jasanoff 2006), the extension to the terrain of human genes shows the constitutive and intimate relationship between the life sciences and what has been termed the bioeconomy and biocapital (Cooper 2008; Pavone and Goven 2017; Sunder Rajan 2006). But as economic historians and sociologists of regulation point out, biocapital significantly depends on the legal instruments and regulatory frameworks that facilitate these economies (Braithwaite and Drahos 2000; Drahos 2016; Mirowski 2011). For legal anthropologists, it became imperative to study how these biolegal developments alter understandings of social life; what are the conditions and circumstances through which culture becomes understood as property?

The *Association for Molecular Pathology v. Myriad Genetics, Inc.* (hereafter *AMP v. Myriad*) case, heard by the US Supreme court and concluded in June 2013, challenged the validity of gene patents.[5] More specifically, the case concerned patents that Myriad Genetics owned on isolated gene sequences used to diagnose a propensity to cancer, in this case breast cancer. Legally, the case revolved around the question of whether human genes (as a general category) were a part of nature and whether or not they were therefore patentable.[6] But at the heart of the debate were questions of access and equality. Proponents of human gene patenting argued that the recognition of such patents would stimulate innovation in genetic research and therefore assist the health of patients. Opponents argued against privatizing what they considered to be public nature, which they believed would in fact stifle research. Moreover, breast cancer patient groups emphasized how patients' choices are curtailed by patents, limiting access to testing, diagnoses, and therapies. Behind the scenes, plaintiffs argued that human gene patents accomplished the opposite of innovation and improvement of health; instead, the patenting of human genes would block patients from accessing genetic information and impede the free study of pathogenetic material that could advance cancer research.

In the end, the Supreme Court ruled that isolated human genes were a product of nature and therefore not patentable. At the same time, however, they also emphasized that *manipulated* human genes *can* be patented. Whereas isolated gene sequences were regarded as products of nature, so-called cDNA—the cloned DNA that in fact makes up the DNA most used in research—was deemed patentable because it involves adequate manipulation. So, while the final ruling of *Myriad* established that isolated human genes could not be patented, seeming to favor the arguments of critics, the verdict also formed new allowances around inventions that involved human (and non-human) genetic material. The ruling, then, was double-edged, recognizing the unpatentability of isolated genes while emphasizing the patentability of altered human genes, shifting the definitional boundaries between nature and invention. To some extent it was a success for the plaintiffs as it allowed human genes to remain in the public domain accessible to anyone who wants to use them for research. Yet, the outcome also allowed biotech companies and private research consortia to continue their business models around cDNA patents. Myriad Genetics and other biotech firms were able to maintain or form monopolies around genetic testing and were incentivized to find novel ways of *manipulating* material to accommodate formal legalities while still appropriating bits of nature.

Following the Supreme Court ruling in the US, debates erupted in Europe, and it was clear that human gene patenting generated more general discussions about the risks of innovation to public morality, ethical practice, and equality. These transatlantic debates exemplified not only the ethico-moral contestations of gene patents, but also the various and conflicting ways innovation and risk itself are deliberated. In pointing to transatlantic differences, scholars show that not only the outcomes, but also the legal process and procedures themselves exemplify different approaches to issues of biotechnology and patents (Hilgartner 2009; Parthasarathy 2017; Jasanoff 2012). Whereas the US follows technocratic experts and allows for a much broader definition of patentable applications, Europe—and the European Commission in particular—operates more democratically in that they invite scientists, activists, and patients into their forums, providing a public voice in the creation of policies.[7] The STS scholar Shobitha Parthasarathy argues that the techno-legal approach from the US ultimately supports market thinking, privileging market actors and imperatives over public policy. In contrast, the EU shows more concern with how the governance of technology via patent systems affects its citizens (Parthasarathy 2017).[8]

Rules around invention make sure that the politics and legalities of patenting in the realm of biology often play out around the process and ethics of invention and these differ by context. Biolegality stresses these differences, not only geographically, in the sense that legal cultures and debates on innovation, invention, and exchange are formed nationally, regionally, or culturally, but also temporally, in that the stakes of property and exchange carry different meanings and values in different historical contexts. These differences also indicate that a case like *AMP v. Myriad*, while demonstrating the particular legal culture of a nation-state at a particular time, also has performative effects beyond its jurisdiction. Because the US still dominates the course of research and development in the global biotech industry, a patent case like *Myriad* structures the course and actions of biotech research worldwide.

Historically, we can say that, as an extension of a sophisticated and enduring system of intellectual property, patents in human genes are expanding into the public domain with "interactive and spiraling effects" (see p. 19 in Winickoff 2015). Such patents make claims on property aspects of human life—social and biological—that previously belonged to the commons. Of course, human gene patents are just one instance of this development. Apart from important strategic investments by private and public sectors, biotech companies in the 1970s were able to flourish because of legal conditions enabling patents to move into the public domain. The *Diamond vs Chakrabarty* case in 1980 was particularly instrumental in this development, allowing, for the first time, the patentability of living matter. In this case, a genetically altered form of bacteria was determined to fulfill the doctrine of "manufacture" or that of "composition of matter," for instance, a newly synthesized chemical compound. This created an important precedent for researchers to patent any such altered micro-organism. A different outcome came with *Harvard College v Canada* (12 May 2002), when the Supreme Court judges ruled, in a 5-4 decision, against the patentability of the Harvard "oncomouse," a genetically altered mouse intended to cultivate oncogenes for cancer research. Our point is that regardless of the differing outcomes of these decisions, patents structure how particular developments in bioscience and biotechnology unfold by shaping the economic terrain research is carried out on. As the bioeconomy gains further traction through public and private incentives and investments, patent cases continue to re-assemble how future R&D agendas are defined. While apparently deciding on the regulation of commercial activity, the Myriad case had a wider implication: It was not just whether nature was patentable, but whether we as humans were patentable. What *Myriad* accomplished was a further fragmentation of nature, but also the fragmentation of humanness.

The patentability of manipulated genetic material has led to, among other things, remarkable inventions of organisms. For instance, the New York Times reports that, as of 2018, there are 13,000 patents involving the DNA of sea life (New York Times 2018).[9] These patents include transgenic organisms invented in research labs by universities or biotech companies. While one cannot patent nature directly—for instance patent your cat—the invention of new transgenic animals using novel techniques and combinations of genes—including those from cats—obtain the status of invention and therefore are eligible for a patenting process.[10] This rearrangement of nature via law is a case *par excellence* of biolegality. In the biolegal process, the status of nature and biology are being reconfigured. The STS scholar David Winickoff describes this process vividly for the *Myriad* case, where "isolated DNA is 'natured' while cDNA, arguably the more financially important assemblage, is denatured in the same breath" (see p. 28 in Winickoff 2015). Both "nature/culture" and "public/private" distinctions are simultaneously reworked in these decisions (see p. 16 in ibid.).

In her book *Patent Politics* (2017), Parthasarathy convincingly shows how patents are more than just legal certifications. As a legal technology, patents have been made compatible and legible across states and national jurisdictions since the beginning of the last century, but their meanings change over time and differ across contexts. Distinct stories accompany these spatiotemporal contexts and while Parthasarathy focuses primarily on the difference between the US and Europe, many other stories emerge when we also engage contexts outside of the Global North (Parry 2004; Foster 2017; Hayden 2003; Pollock 2019), as we will show later in this chapter.

A leading justification for patents across contexts is that of invention and innovation. Patents are widely thought to enhance research and aid the progression of biomedical applications such as new drugs and therapies. But while invention forms the central theoretical premise of modern patent law, studies by social scientists have shown that patents "do more than influence innovation trajectories and business strategies":

> As they explicitly define patentable 'invention,' they are also shaping social understandings of nonpatentable 'discovery' and 'nature.' In helping to generate markets for new technologies, they emphasize particular kinds of economic relationships at the expense of others. Finally, researchers have found that the descriptions contained within the patents themselves can shape how we define social categories such as race and citizenship. (see p. 4 in Parthasarathy 2017)

Important for our idea of biolegality, this statement makes clear that patents are more than just legal technologies, producing effects in the social and economic space as much as they do in legal and knowledge domains. While Parthasarthy talks about "patent politics," we would add that besides the politics, it is the moral weight, the technical specificity, and the mobility of law by which patents become not only political lightning rods across jurisdictions, but also objects in their own rights that can be traded in the (global) bioeconomy.

The latter has been theorized by the legal scholar Hyo Yoon Kang (2012, 2020) who sought to examine patents as legal techniques. In her analysis of the material infrastructure that sustains the contemporary patent, she shows how the patent becomes a financialized asset, to be traded individually or as part of a portfolio. This *assetization* of patents—or intellectual property rights more broadly—does not merely commodify the original invention but turns law itself into a speculative financial asset. It does so by enacting patents "as real options in valuation practices [that] have been used as instruments of financial hedging" (see p. 50 in Kang 2020). Here, patents are involved in the speculation of financial risk, orienting profits into the future and thereby actively shaping biomedical developments. In this scenario, *promissory* capital is traded, and the focus is less on sale profits. As the sociologist Victor Roy (2020) notes, "this ontology of assets not only privileges certain kinds of assets—those that can generate continual accumulation—but it is also intertwined with changes in our understanding of health and illness" (see p. 99 in Roy 2020).

More recently, Biagioli and Pottage have examined how the turn to bioinformation has changed patent law: "[B]ioinformation abstracted from the body is returned to the body as a prompt for the body's own information processing circuitry… It is this shift from an industrial or manufacturing paradigm to a bioinformational paradigm that informs the new sense of invention" (Biagioli and Pottage 2021: 233). Here, patent law is troubled by developments that stress biological processes in silico rather than in vitro, upsetting the figure of nature (discovery) central to law's demarcation from culture (invention). External to legal technique, analysis of patent law in different geographical contexts emphasizes cultural reconfiguration. Against the straightforward neoliberal argument that law facilitates biocapital in globalization—using biomaterials or bioinformation as currency—these studies emphasize ethical and translational work in the appropriation of patents (in and outside of the Euro-American context) that give a more nuanced understanding of exchange, one that does not preclude neoliberal forces but that recognizes alternative forces negotiating with them (Foster 2017; Parry 2002; Strathern 2001). Along these lines, biolegality prompt new modes of exchange: economic, scientific, and cultural.

Moving away from patents, what we would like to describe next is how we see biolegality operating when individuals—and not companies, universities, or scientists—claim ownership. Here, the subject of ownership or property over one's own body or bodily material finds new traction in a postgenomic world.

Bodily Ownership: From Waste to Genetic Personhood

The growing collaboration between markets and biomedical technologies in the 1980s—primarily in the Global North but increasingly in China, East Asia, and South Asia—led to an increased need for human biological material including, though not limited to, human skin, bones, heart valves, embryos, eggs, semen, stool samples, stem cell lines, and more generally, genetic material and genomic information. Hence, the globalized bioeconomy depended on the legal circulation of such material, making law central to the mechanisms of biocapital (Sunder Rajan 2006; Mirowski 2011). While the previous section gave an account of how these economic interests influenced disputes around human gene patents, the focus here is to see how bodily material in general, and genetic information in particular, are caught up in debates around ownership and personhood. While there are many cases highlighting the ways in which self-ownership over bodily material troubles notions of personhood, we single out two cases prominently discussed in the scholarship: the case around the immortal cell line of Henrietta Lacks and the landmark court case *Moore v. Regents of the University of California*

(1990, the *Moore* case hereafter) about the uses of John Moore's spleen for commercial purposes.

Although these cases do not explicitly engage with the genetic or the genomic, we demonstrate how the connection between bodily material and personhood is given new meaning in a genomic age. As scholars have shown, regardless whether genetic material is isolated, modified, altered, or "edited," unless grown from cultured cells (as with cloning), the origins often start from a body (through a swab, prick, or biopsy) and remains linked to the body through language and representation (Everett 2007; Parry 2004; Strathern 2005). The boundaries drawn between material that belongs to, or is released from, the body are played out most vividly in law, particularly via debates around personhood. The representational power of genetics in contemporary Euro-American societies—exemplified by the socialization of genetics as the "code of life" and by coupling genetic material to identity—revalued (human) biological material such that, for instance, what was previously "biological waste," legally speaking, is redefined as "genetic material." Understanding the *Moore* case within this new postgenomic context allows us to reconceptualize the relationship between biological material, genes, and persons. This is also true for understanding the discussion about the cells of Henrietta Lacks, which never made it to court, but stands as an important case that ignited heated debate in the US and beyond. Let us start with this controversy, which, when analyzing the debate's temporal impact in the public domain is exceptional.

In the 1950s, researchers at the Johns Hopkins University Hospital were working on making immortal cell lines, namely modifying cells so that they can be cultured indefinitely. Immortal cells simplify laboratory research and make it much more cost-effective. At one point they obtained cells from Henrietta Lacks, who suffered from a rare form of cervical cancer, and these cells multiplied aggressively, making it possible to mass produce what became known as HeLa cells. HeLa cells are ubiquitous in medical research labs across the world, where they are used to further cell culture techniques, explore chromosomal abnormalities, the functioning of viruses, and cervical cancer research more generally (see p. 3 in Quigley 2018). The HeLa cells only became controversial in the US following publication, in 2011, of science writer Rebecca

Skloot's *The Immortal Life of Henrietta Lacks*. The book detailed the story of Henrietta Lacks and the legacy of her cells, becoming a New York Times number one bestseller. An African American woman, and mother of five, Henrietta Lacks—born and raised in rural Virginia—lived in poverty in the heart of the Jim Crow South, while her immortal cells, following her death, became increasingly valuable—and helped create even more value—through the public and increasingly commercialized practices of biomedical research. Public controversy around consent and privacy rights erupted following the popularity of Skloot's book and in 2013 the National Institute of Health (NIH) returned control over access to the HeLa cell lines' data to the Lacks family.

Historian of science, Hannah Landecker, examined the shifting meanings of the relationship between the HeLa cells and the person of Henrietta Lacks, explaining how the personification changed depending on historical context. While the 1950s linked the biomedical miracle and wide distribution of the HeLa cells to the image of a "beneficent young Baltimore housewife," the 1960s discovery of the contaminating potential of HeLa cells reframed them as the product of a "promiscuous, malicious black woman" (see pp. 212–213 in Landecker 1999). Whether charitable or spiteful images represented the personification of HeLa cells, "the cells were seen as physically separate yet continues in essence with the body and the person of the woman they came from" (see p. 213 in Landecker 1999).

Meanwhile, the 1980s brought significant changes to the understanding of biological material as a form of property. The emergence and consolidation of patents for recombinant DNA changed the way researchers shared their biological material; while this facilitated the commodification of human biological material, the patenting system also altered legal thinking around property and ownership. These formed new biolegal conditions in which to place Skloot's book and the subsequent scandal fifty years after Henrietta Lacks' death. But it is important to acknowledge that part of the biolegal regime is also the process of personification described so eloquently by Landecker (ibid.; see also p. 16 in Lee 2012). This process interpellated the history of US racism embodied by chattel slavery, medical experimentation on racialized women, and racial science in general (see also hooks 2012; Landecker 1999; M'charek

2021). It is also important to note that Skloot's book entered US public consciousness at the height of racial tensions following the murder of Trayvon Martin in 2012 and the subsequent exoneration of his shooter. In fact, in their legal effort in 2021 to seek compensation from pharmaceutical companies, the Lacks family hired Ben Crump, the well-known civil rights attorney who also represented the families of Trayvon Martin, George Floyd, and Breonna Taylor.

As the Henrietta Lacks example shows, there is no need for an actual court case to invoke a sense of right, legality, and legitimacy. Skloot's book was a bestseller and reached the status of mandatory reading for many college and university bioethics courses in the US. The subsequent debate led the NIH and the Obama administration to grant her descendants some control over the use of HeLa lines and to include them in discussions around the future of biomedicine and patient rights. The Lacks' family did file a lawsuit against Thermo Fisher Scientific in 2021, but this was settled out of court.[11] Although symbolically, Henrietta Lacks' contribution to science was repaired and her rights acknowledged by powerful institutions, there was never a legal ruling granting ownership of one's own biological material. This was because the HeLa controversy came in the wake of *Moore v. Regents of the University of California*, which delivered the defining legal statement that one's own human biological material cannot be regarded as one's own personal property, and consequently, that individuals have no claim on profits yielded from products or research derived from their cells.

The *Moore* case concerned an Alaskan man, John Moore, who, in 1990, sued his former physician for using his cells for research and commercialization without prior consent. Suffering from a rare form of leukemia in 1976, Moore's spleen was used (along with other tissue) by his treating physician Dr. Golde to create and patent (together with the UCLA Medical Center) the so-called "Mo cell line" (later renamed "RLC"). This happened between 1976 and 1983 and during this time Dr. Golde never disclosed to Mr Moore any financial interest in the biological materials that were taken from him. While the Court of Appeal in California agreed with Moore's claim that "he continued to own his cells following their removal from his body,"[12] the Supreme Court of California overturned the ruling by arguing that "imposing a liability in conversion on researchers could hamper research and its potential benefits to society."[13] The case was important because the California Supreme Court did not regard samples from discarded blood and tissue as personal property, which had direct impact on all subsequent possible litigants who were seeking compensation for the improper use of their deceased family members' human biological material (including the Lacks family). At its core, the Supreme Court ruling was framed in terms of intellectual property and in the name of future innovations. Even though the court also decided that Moore could bring an injury claim (for the fact that Dr. Golde failed to disclose his financial interests), this would have been impossible to realize, since Moore would have to give evidence that he—or any "reasonable person"—would not have consented to the proposed splenectomy, a surgery he needed to continue his cancer treatment.[14]

Cases like Henrietta Lacks and *Moore* have been thoroughly discussed across disciplines in the social sciences, humanities, and law (see Frow 1997; Rabinow 1999; Hayden 2007; Hoeyer 2013). As these accounts illustrate the extent to which individuals feel their biological material still belongs to them even as it is separated from their bodies, these instances also demonstrate the affects of increased commercialization of bioscience and biomedicine on the changing characteristics of law. Discussing the category of "waste," Waldby and Mitchell argue that the tissue economies embodied by cases like *Moore* exemplify a new situation, different from the tissue economies attached to organs and blood, which was based on reciprocal exchange and pre-known or stable uses (see p. 108 in Waldby and Mitchell 2006). Waldby and Mitchell theorized the economies of labor characterizing the new situation, where the removal, storage, and transport of transplant tissues became fully commodified. Moreover, tissue economies became increasingly based on speculative social ecologies; Moore's cells were only valuable as potential speculative assets insofar as a diseased spleen was first and foremost seen as waste: "In their role as constitutive waste, Moore's donated (rather than owned) tissues were part of the basis for the fiscal and biological health of the body politic" (see p. 109 in ibid., 109).

Emerging research on the human microbiome exemplifies another development of potential value in waste, in this case human fecal waste.[15] For instance, Fecal Microbiota Transplants (FMT) have proven very successful for the treatment of Clostridium difficile infection.[16] While FMT treatments are growing rapidly throughout the world and forming public stool banks is becoming standard practice, the politics of legal categorization, confronts the question of expanding or limiting the circulation of feces and microbes, including furthering or limiting the possibilities of commercialization.[17] Similar to the spleen cells in the *Moore* case, stool flows through different regimes of value, making the dynamic legal process constitutive to the production of commercial and clinical value. While legal cases have yet to arise out of these new practices, ethical discussions are being held about the status of microbiota—namely whether they should be considered as part of the human body. Microbiota are separate, diverse living organisms living inside of us and do not share our genetic material, yet they can very well be regarded as constituting part of our biological self. Research in the human microbiome ultimately suggests that the distinction between what is and what is not human becomes increasingly blurred. This situation complicates our social understanding of self and legal understandings of personhood. Microbiota can also be seen as a product of the human body, implicating the research within the labor theory of property. This complicates any ethical measures in both research and commercial application.

From the standpoint of biolegality, the following two issues are at stake. First, as scholars inside and outside of legal theory argue, biomedical advances are weakening existing legal tools for marking the boundaries of personhood (Jasanoff 2016; Hoeyer and Hogle 2014; Radin 2018). For instance, the classic boundary-making device "informed consent," which has already been challenged as inadequate in the context of the different perspectives of non-western cultures and ontologies (Simpson 2013), is also weakening because of personhood's fragmentation in the age of biotechnology. In the case of Henrietta Lacks and the NIH, one could argue that the ensuing agreement between the NIH and Lacks' family represents an acknowledgment of the failure to secure informed consent and is an attempt to restore the interests of Lacks' family by granting their rights to determine access HeLa cells for scientific research. But the marker for personhood is distributed over family members and the agreement is a reparative one-off gesture forged between an institution and the family. Francis Collins, NIH's director responsible for the deal at the time, forestalled any claim to precedent in adjudicating other cases involving human biological material (Callaway 2013). The uniqueness of this case—a singularity rarely afforded in law and legal practice where generalizability is the prerequisite for equality before law—evinces a biolegal configuration: while the solution takes place outside of the conventional legal system, it invokes and mediates new expectations, processes, and standards that rearrange classical devices such as informed consent.

Informed consent, which has seemed like the logical instrument to secure ethical process, is becoming more and more insufficient for securing personal autonomy and control in the face of the concentration, quantity, changing nature, and complexity of data (that can have serious health implications for self and kin). As the *Moore* case shows, devices such as informed consent become unstable when combined with other more implicit tools of regulation and law, such as definitions and categories. For instance, Waldby and Mitchell refer to the constitutive nature of the category "waste" in establishing the biovalue of Moore's spleen cells. In and of itself, his cells were useless; it needed a "speculative social ecology" (see p. 109 in ibid.) to make them valuable. Bearing in mind the accumulation of such speculative social ecologies under the umbrella of innovation, and the promissory narratives of hope and vast benefits following these developments (Taussig, Hoeyer and Helmreich 2013), the ways researchers produce biovalue from otherwise "useless" material becomes ever more important. Moreover, as Landecker so powerfully argues regarding the cases of HeLa and Mo lines, value is not only produced by the status of the material, in this case a cell line, but for the *reproductive* potential that such unique cell lines have (see p. 222 in Landecker 1999). HeLa and Mo cells were able to be capitalized because of their capacity to reproduce more of itself. It is this reproductive characteristic of such biologicals that complicate matters pertaining to law and regulation. These complications are further highlighted by bioscience's shift from using biologicals for research to only needing bioinformation; the change from in vitro to in silico analysis.

The Challenge of Bioinformation and Genetic Sequence Data

Developments in gene therapy and precision medicine indicate that genomic sequencing is becoming standard practice in clinical and scientific spaces. Faster and more economical sequencing technologies have significantly contributed to this development.[18] These new methods of genomic sequencing have made biochemical information and genetic data more readily "accessible, transmissible, and replicable, than it was when embedded in whole organisms" (Parry 2004: 201), ushering in a shift from biomaterials to bio*information*: information stemming from biological organisms and materials. With the life sciences' sine qua non relying more and more on bioinformation rather than biomaterials, anthropologists, philosophers, geographers, and legal theorists are now studying its consequences for law and regulation (Gere and Parry 2006; Leonelli 2010; Parry 2004; Parry and Gere 2006; Parry and Greenhough 2017).

According to Parry (2004), the shift in biotechnology to relying more on proxies and bioinformation—cell lines, frozen tissue samples, genetic sequence data, etc.—parallels earlier developments in informational technologies, where books, music, and movies—media formerly contained on material substrates and accessed physically via purchase or rental at brick-and-mortar shops (book/record/video-rental shops)—are now primarily digital and shared, transmitted, copied, and replicated via the internet. As she remarks: "just as new informational technologies have enabled particular resources to be structured in ways that have made them easier to circulate, store, and reprocess, so biotechnology has also enabled whole organisms to be rendered in ways that make the genetic and biochemical materials and information embodied within them much easier to transmit, store, reprocess, and recirculate" (see p. 250 in Parry 2004). The increasing preference of the life sciences for bioinformation rather than material forms poses regulatory challenges: how to assess ownership and property claims when the commodity at hand is intangible and infinitely reproducible information that is considerably more fungible—and piratable—than the circulation of whole organisms?

As lawmakers ponder how to regulate bioinformation as immaterial "material," Parry and Greenhough assert that bioinformation should not be seen as either material or immaterial, but instead, as "*materialized* in different ways at different points in existence" (see p. 8 in Parry and Greenhough 2017). They distinguish between derivative and descriptive bioinformation, where the first derives directly from the organism—and is "fleshier"—and the second comprises narrative descriptions about individuals—and is "wordier" (see p. 5 in ibid.).[19] According to them, bioinformation is never static but "always relational," dependent upon who uses it and for what purpose (see p. 7 in ibid.). Such a situated understanding of bioinformation also affects how we analyze its implications for commercialization.

This is exemplified in the 2000s debate around the intellectual property rights over genetic sequence data (GSD) between the National Institute for Health (NIH), in charge of the public Human Genome Project, and the US-based biotech firm Celera Genomics led by the well-known entrepreneur Craig Venter. Celera Genomics claimed IP rights over a selection of data, arguing that their shotgun sequencing technology that produced these data was much more effective than others, thereby justifying their prerogative to balance their investment with exclusive rights. The company charged a fee for early access to its sequence data and went on to patent human genes to further their biomedical research. However, arguments by publicly funded scientists for making genetic data public were much stronger. Earlier, these ethical standpoints had led the largest labs involved in the human genome project to create the so-called "Bermuda Principles." Formulated by leading scientists of the Human Genome Project in 1996, the Bermuda Principles (or "Bermuda Accord") contain guidelines for releasing genetic sequence data to the public. Rather than releasing data after publication, the rules stipulated that all genetic sequence data are to be released within twenty-four hours after they have been generated. These principles were shapeshifting, and its norms were adopted by many institutions. While the Bermuda Accord was successful in socializing and sustaining the virtue of open access, challenges like those presented by Celera Genomics did occur. Despite their own policies, journals sometimes made concessions with companies to allow non-open access sequence databases (Lawson and Rourke

2016). Notwithstanding these exceptions, open science principles remain the dominant trend.

A strong open data governance movement has emerged in the past decade. Defending the principles of "Open Science," where the sharing of data, results, and methods is regarded as core to successful and ethical scientific collaboration, the governance of open data by scientists, governments, and research institutions receives broad institutional endorsement. Formally adopted by the European Commission in 2016, principles of Open Science were authorized by the biggest funding agencies in Europe and the US, including the NSF, the Welcome Trust, and the Bill and Melinda Gates Foundation (see pp. 5–6 in Beaulieu and Leonelli 2021). These developments create a research culture that prioritizes the sharing of genetic sequence data, a culture epitomized during the recent COVID-19 pandemic when this ethos enabled sharing practices amongst Asian scientists and their counterparts in the rest of the world, disseminating the virus' genetic code so that research on them could start as early as possible, accelerating the rate that COVID-19 vaccines were created (Fearnley 2020). But this ethos stands in stark contrast to private uses of Open Data for commercial entities on the one hand, and, on the other, to debates about the data sovereignty of (indigenous) communities or nation-states. Both complicate rules around global scientific exchange.

Genomic Science and Its Exchanges Across Borders

The questions around property and exchange raised by genomic technologies are compounded by the effects of globalization. With respect to the circulation of bioinformation, EJ Gonzalez-Polledo and Silvia Posocco rightly point out:

> In the age of global infrastructures, bioinformation is routinely donated, exchanged, traded or extracted under diverse arrangements and transactional forms that involve persons, institutions and biotechnologies.

> Indeed, the global reach of bioinformation infrastructures make bioinformation a "global social fact" – a form of relationality that links multiple localities, infrastructures, scientific cultures and interests. (See p. 2 in Gonzalez-Polledo and Posocco 2021)

Law and regulation play an integral part in controlling (or "disciplining," as we explain further along) biological material and information in these global infrastructures: while some regulations, on the one hand, facilitate industry and commerce, others are intended to mitigate inequalities generated through the globalization of biomedicine and bioscience. Like biotechnology, globalization elicits "new conceptions of the functioning of legal institutions" (see p. 8 in Pottage 2004).

In *Trading the Genome*, Bronwyn Parry (see p. 240 in 2004) examines these global changes and argues that in the (colonial) past biomaterials were "disciplined" by the epistemological categorization of living organisms, especially in the eighteenth-century Linnaean classification system. In this way, biomaterials were made legible and familiar across different geographies. But in late capitalism, Parry argues "the Euro-American system of intellectual property rights law" represents a new way of categorizing biological material and information (see p. 240 in ibid.). Ironically, in the past, wealth was created by circulating whole plants and organisms, while, starting in the 1990s with what has been termed bioprospecting, it is control over the means of circulation that generates income and influence.

In addition to the science, bioprospecting was made possible through the global application of intellectual property rights instituted by the 1995 Agreement on Trade-Related Aspects of Intellectual Property of the World Trade Organization, better known as TRIPS. There is a solid body of scholarship in the field of governance and international relations describing how TRIPS—in a relatively short time—has significantly altered the global space of trade arrangements (Drahos and Braithwaite 2001; Sell 2003). According to historian of science, Philip Mirowski, this global intellectual property system privatized science and did so on the terms that were set out, predominantly, by the US and to a lesser extent, the EU and Japan (see p. 186 in 2011).

Sustained resistance from postcolonial nation-states and indigenous communities throughout the TRIPS process has pushed the need for regulatory frameworks that would mitigate the unequal distributions of wealth extracted from these new trade arrangements. Within the biosciences, the inequality problem included the enormous profits generated by the pharmaceutical industry's bioprospecting ventures in Global South countries where the communities and nation-states providing the biological material receive no remuneration. The 1994 Convention on Biological Diversity (CBD) offered one formal avenue for contractual negotiations between universities and pharmaceutical companies in the Global North and researchers and their institutions in the Global South around the exchange of biological resources following the principles of so-called benefit-sharing principles.

While most studies on regulation and biotechnology conducted outside of the Global North have focused on bioprospecting, or how plants are extracted for commercial purposes by pharmaceutical industries (Foster 2017; Hayden 2003; Parry 2004), genomic sequencing technologies expand these practices to the domain of humans (see p. 15 in Reardon 2009). Following the human genome project, nation-states are increasingly caught between demands to strengthen their economies and to secure their biological resources in response to the global surge in genomic research (Rabinow 2011; Hayden 2003; Mitchell and Waldby 2006). Global South countries which supply biological samples used by researchers and pharmaceutical companies in the Global North who then claim intellectual property rights over the—expensive and profitable—products are most affected and disadvantaged in the flurry of genomic research (Irwin 2010). Global South states defensively respond by asserting what scholars describe as forms of "biosovereignty" (see p. 40 in Ong 2010, and pp. 60–62 in Ong 2016, Hinterberger and Porter 2015; Hurlbut 2017) in an attempt to protect their biological resources from global market forces.

Hence, biolegal claims get more complicated as multiple jurisdictions are involved. In the last few decades the exchange of biological materials has become more ethically fraught with the rise of genomic sciences globally, and the growing power of global pharmaceutical companies. In order to study biological materials indigenous to their communities, scientists in the Global South depended on scientists in the Global North, who have access to better laboratories and institutional resources. Biolegal configurations are premised on postcolonial power relations (Anderson 2000; Kowal 2013; Kowal et al. 2013; Kowal and Radin 2015), where imperial and colonial histories continue to reverberate in the present, not least in relationships between researchers.

Some nation-states have worked to secure national genetic databanks to assert themselves in terms of nation building and to seek new domestic revenues. National biobanking aims to produce biovalue, but also creates new forms of clinical labor that alter medical and popular understandings of risk for common diseases and conditions (Mitchell and Waldby 2010). At stake then is not only the nation-state's sovereign power to protect and/or capitalize on their genetic resources, (genetic) biobanking also represents the neoliberalization of national health care systems, providing industry with the resources to commercialize risk analysis (via, for instance, DNA testing) and preventative medicine (through precision treatments and other forms of targeted health care). Within a settler-colonial, regional, or global context, forms of biobanking do not merely engage global science, they are also new sites for asserting national identities (Schwartz-Marín and Méndez 2012; Slabbert and Pepper 2010), indigenous rights (Kowal 2013; Reardon and Tallbear 2012), and postcolonial sentiments (Benjamin 2009). While regulatory frameworks are key to facilitating biobanks in national jurisdictions, states and indigenous communities call upon legal instruments to mitigate colonial and postcolonial inequalities generated through the globalization of biomedicine and bioscience.

Conclusion

This chapter engaged the broad question of how biotechnology is dramatically altering the way twenty-first-century society and its institutions produce and revise the rules and conventions around ownership, property, exchange, and sharing. The alterations are not merely the subject of legal reform; they are in essence anthropological and philosophical concerns. The gene—as biological material, bioinformatic sequence, patentable technological object, and speculative intellectual property—is an essential object for thinking through and capturing these changes. It is important to place these bioscientific developments within the context of the growing bioeconomy (Pavone and Goven 2017; Sunder Rajan 2017). With the current increased focus on, and development of, bioinformatics, ehealth services, mobile health data, private biobanks, commercial genetic testing, all aiming to form the basis for personalized or "precision" medicine, the question of ownership and personhood is in great disarray. The contemporary individual is fragmented not only through immortal cell lines, but also through immortal data that further complicate rules and conventions around persons, bodies, and property.

The debates and practices around the gene and genome exemplify the changing thinking around the meaning of invention, the manipulation of nature, and its incorporation into culture. These advances, coupled with the proliferation of intellectual property rights, have also changed how we, as individuals and communities, think differently about how our bodies, our (biological or clinical) labor, our "selves" as persons, and our bioinformation relate to the question of property and ownership. The ontological changes to legal instruments (through the assetization of patents) and biological matter (through datafication) demonstrate the operation of biolegality, revealing not only the malleability of biological materials, but the malleability of law itself. Finally, all these changes are intensified by the process and practice of globalization. In a time where nation-states claim sovereignty over their genetic resources, global rules for scientific exchange continue to privilege the protection of intellectual property rights asserted primarily by actors in the Global North. At stake then, when considering the social life of the genome, is not only the scientific knowledge emanating from its analysis, but also how (global) society (re)organizes the wealth and value that emanates from collecting, using, and manipulating this knowledge.

Notes

1. Thanks to Ben Stork for his acute observation about the movement of metaphor to synecdoche.
2. Differences are on the level of national politics, as in terms of economics we can argue that institutions such as the WIPO make patenting practices global. More on this later.
3. In discussing legal personhood, the chapter necessarily spills over in the next one. Debates on property are inherently entangled with personhood, but whereas the stress on genes in this chapter will be on the relationship between persons and things, the next chapter will mostly cover the relationship between personhood and responsibility.
4. Of course the idea that the (genetic) code underlying biology was more important than the material itself was long present since the structure of DNA was established. But before high-throughput sequencing it was always necessary to bring material to the lab. Without this requirement—as mobile sequencing can happen anywhere—ownership becomes directly coupled with code rather than the material.
5. *Association for Molecular Pathology v. Myriad Genetics, Inc.*, 569 U.S. 576 (2013). To give a brief context of the *Myriad* case, the gene patents that were granted to Myriad Genetics involved both cDNA and the "isolated DNA sequences" and were granted in 1994. This was Myriad's core business, and the BRCA breast cancer test was developed by way of these patents and hit the consumer market in 1996. However, patient groups and researchers started to organize themselves after it became clear that Myriad Genetics—but also other holders of human gene patents—acquired monopoly statuses on genetic testing, making it cost-prohibitive for patients and researchers. These practices eventually led the American Civil Liberties Union (ACLU) and Public Patent Foundation to file a suit against Myriad Genetics in 2009. Two lower federal courts were involved before it came to the Supreme Court. For a fuller history of the BRCA gene patents see Winickoff (2015).

6. The central "product of nature" doctrine originates in the nineteenth century and stipulates that an invention cannot include "products of nature," including "laws of nature, natural phenomena, and abstract ideas" (see p. 17 in Winickoff 2015). The "product of nature" doctrine was strategically central to the plaintiffs' strategy since the actual concerns related to equal access to research and health care was deemed less viable for success. See Parthasarathy (2017) for an elaborate discussion.
7. The more recent debate around the regulation of genetically modified food and the definition of "genetic modification" is telling here; whereas the US focuses on the end product and understands modification as that which has altered genetic material, European definitions focus on the process and include in the definition the process of altering, whether or not it has altered the end product.
8. It might also be useful here to point out the difference with the Australian context where the Australian High Court in *D'Arcy v Myriad Genetics* not only ruled that isolated genes were unpatentable but also that cDNA (or any naturally occurring substance that contains genetic information) is excluded as patentable. Providing a clear taxonomy of what was deemed patentable and not patentable, the Australian High Court was seen as delivering a more decisive outcome and effective set of guidelines.
9. https://www.nytimes.com/2018/09/17/science/patents-marine-dna.html.
10. Or sometimes the status of original creative works and thus can be copyrighted.
11. More recently—just before submitting this manuscript—the Lacks family settled a lawsuit against the biotech company Thermo Fisher Scientific (see https://www.nytimes.com/2023/08/01/science/henrietta-lacks-cells-lawsuit-settlement.html). While claims of ownership were legally formalized via the filing of a lawsuit, the settlement prevented any presentation of or decision on the ownership claims thus rendering no legal precedent to recognize cells as one's property.

12. Moore v. Regents of the University of California 249 Cal. Rptr. 494 at 508 [1988] at 508 [Moore, Court of Appeal].
13. Moore v. Regents of the University of California 51 Cal.3d 120 at 127 [Cal, 1990] [Moore, Supreme Court]. In legal practice, the use of "conversion" refers to the unlawful use of one's property.
14. The "reasonable person clause" is part of the disclosure obligation. Here the information material by the physician to the patient is only considered material if "a reasonable person" would share the importance of the risks in considering a proposed therapy (see p. 572 in Leeds 1991).
15. This research indicates that microbes do not merely live in and on our bodies for their own survival; they extract energy from food, they control the immune system, and prevent pathogens from multiplying. Hence, the promise of microbiome research lies in these functions, for instance the ways drugs are metabolized by the microbiome or the ways in which transplants of certain microbiota can disengage pathogens, such as bowel infections.
16. The treatment originates from a Dutch study conducted in the year 2000 with 400 people suffering from C-diff and proved extremely successful. Results indicated a primary and secondary cure rates of over 85% compared to the 20% cure rates with antibiotics. In fact, the study was so successful that it was terminated for ethical reasons. The ethical board found it unethical that c-diff sufferers in the control group were not given the same treatment.
17. In 2013, the Food and Drug Administration in the US determined that stool for the purposes of FMT should be regulated as a drug without any approved indication (Cohen 2019). This means that medical doctors wanting to perform FMT had to register the procedure as part of a clinical trial or submit an "Investigational New Drug" (IND), both of which are known to be very bureaucratically onerous, taking months to complete. These rules were recently mitigated for C-diff treatments and physicians are now allowed to perform FMTs without INDs.

18. Increasingly, high throughput sequencing needs only data as methods used in major genome sequencing are the clone-by-clone method and the whole genome shotgun sequencing. Shotgun sequencing involves "shearing genomic DNA into short nucleotide strands, cloning these strands to create a DNA library, and then using massive computing power to recompose the entire human genome by identifying and marrying overlapping sequences" (see p. 23 in Pottage 2007).
19. While derivative bioinformation could refer for instance to genetic data from human tissues, descriptive bioinformation designates information about individual's lives, such as disease experience, mortality risk or social identity (see p. 5 in Parry and Greenhough 2017).

References

Anderson, W. 2000. The possession of kuru: medical science and biocolonial exchange. *Comparative Studies in Society and History*, 42(4), pp. 713–744.

Beaulieu, A., & Leonelli, S. 2021. *Data and Society: A Critical Introduction.* SAGE.

Benjamin, Ruha. 2009. A Lab of Their Own: Genomic Sovereignty as Postcolonial Science Policy. *Policy and Society*, 28(4), pp. 341–355.

Biagioli, M. 2006. Patent Republic: Representing Inventions, Constructing Rights and Authors. *Social Research: An International Quarterly*, 73(4), pp. 1129–1172.

Biagioli, M. 2011. Patent Specification and Political Representation: How Patents Became Rights. *Making and Unmaking Intellectual Property: Creative Production in Legal and Cultural Perspective*, pp. 25–40. Chicago: University of Chicago Press.

Biagioli, M. and Pottage, A., 2021. Patenting Personalized Medicine: Molecules, Information, and the Body. *Osiris*, 36(1), pp. 221–240.

Braithwaite, J. and Drahos, P., 2000. *Global Business Regulation.* Cambridge university press.

Carolan, M. S. 2010. The Mutability of Biotechnology Patents: From Unwieldy Products of Nature to Independent 'Object/s'. *Theory, culture & society*, 27(1), pp. 110–129.

Callaway, E. 2013. Deal Done Over Hela Cell Line. *Nature*, 500(7461), pp. 132–133.

Cohen, M. 2019. The Law of Placenta. *Yale Journal of Law and Feminism*, 31, p. 337.

Cooper, M. E., 2008. *Life as surplus: Biotechnology and capitalism in the neoliberal era*. University of Washington Press.

Drahos, Peter and Braithwaite, John 2001. Intellectual Property, Corporate Strategy, Globalisation: TRIPS in Context. *Wisconsin International Law Journal*, 20, p. 451.

Everett, M. 2007. The 'i' in the Gene: Divided Property, Fragmented Personhood, and the Making of a Genetic Privacy Law. *American Ethnologist*, 34(2), pp. 375–386.

Fearnley, Lyle. 2020. Viral Sovereignty or Sequence Etiquette? Asian Science, Open Data, and Knowledge Control in Global Virus Surveillance. *East Asian Science, Technology and Society: An International Journal*, 14(3), pp. 479–505.

Finkler, K. 2000. *Experiencing the New Genetics: Family and Kinship on the Medical Frontier*. Philadelphia: University of Pennsylvania Press.

Foster, L. A., 2017. *Reinventing Hoodia: Peoples, Plants, and Patents in South Africa*. University of Washington Press.

Gere, Cathy, and Parry, Bronwyn. 2006. The Flesh Made Word: Banking the Body in the Age of Information. *BioSocieties*, 1(1), pp. 41–54.

Gonzalez-Polledo, Elena J. and Posocco, Silvia, eds. 2021. *Worlds and Futures*. New York and London: Routledge.

Haraway, Donna 1997. *Modest _Witness @Second_Millenium.FemaleMan_ Meets_OncoMouse™: Feminism and Technoscience*. New York: Routledge.

Hayden, Cori. 2003. *When Nature Goes Public. The Making and Unmaking of Bioprospecting in Mexico*. Princeton and Oxford: Princeton University Press.

Hayden, C. 2007. Taking as Giving: Bioscience, Exchange, and the Politics of Benefit Sharing. *Social Studies of Science*, 37(5), pp. 729–758.

Hilgartner, Stephen. 2009. Intellectual Property and the Politics of Emerging Technology: Inventors, Citizens, and Powers to Shape to the Future. *Chicago-Kent Law Review*, 84(1), pp. 197–226.

Hinterberger, A. and Porter, N., 2015. Genomic and viral sovereignty: Tethering the materials of global biomedicine. *Public Culture*, 27(2), pp. 361–386.

Hoeyer, K. 2013. *Exchanging Human Bodily Material: Rethinking Bodies and Markets*. Dordrecht: Springer Science & Business Media.
Hoeyer, K. and Hogle, L.F., 2014. Informed Consent: the Politics of Intent and Practice in Medical Research Ethics. *Annual Review of Anthropology*, 43, pp. 347–362.
hooks, b. 2012. *Writing Beyond Race: Living Theory and Practice*. London: Routledge.
Hurlbut, J. B. 2017. A science that knows no country: Pandemic preparedness, global risk, sovereign science. *Big Data & Society*, 4(2), p. 1–14.
Jasanoff, S. 2006. Just Evidence: The Limits of Science in the Legal Process. *The Journal of Law, Medicine & Ethics*, 34(2), pp. 328–341.
Jasanoff, S., 2012. *Science and Public Reason*. Routledge.
Jasanoff, S. 2016. Perfecting the Human: Posthuman Imaginaries and Technologies of Reason. In *Perfecting Human Futures*, pp. 73–95. Wiesbaden: Springer VS.
Kang, H. Y. 2012. 'Science Inside Law: The Making of a New Patent Class in the International Patent Classification', *Science in Context*, 25(4), 551–194.
Kang, H. Y. 2020. Patents as Assets: Intellectual Property Rights as Market Subjects and Objects. In Birch, K. and Muniesa, F. (eds.), *Assetization: Turning Things into Assets in Technoscientific Capitalism*, pp. 45–74. Cambridge: MIT Press.
Kowal, E. 2013. Orphan DNA: Indigenous samples, ethical biovalue and postcolonial science. *Social Studies of Science*, 43(4), pp. 577–597.
Kowal, E. and Radin, J. 2015. Indigenous Biospecimen Collections and the Cryopolitics of Frozen Life. *Journal of Sociology*, 51(1), pp. 63–80.
Kowal, E., Radin, J. and Reardon, J. 2013. Indigenous body parts, mutating temporalities, and the half-lives of postcolonial technoscience. *Social Studies of Science*, 43(4), pp. 465–483.
Landecker, H. 1999. Between Beneficence and Chattel: the Human Biological in Law and Science. *Science in Context*, 12(1), pp. 203–225.
Lawson, C. and Rourke, M. 2016. Open Access DNA, RNA and Amino Acid Sequences: The Consequences and Solutions for the International Regulation of Access and Benefit Sharing. *Griffith University Law School Research Paper* (16-12).
Lee, C., 2012. The Unspoken Significance of Gender in Constructing Kinship, Race, and Nation. In Wailoo, K., Nelson, A. And Lee, C. Eds., Genetics and the Unsettled Past: the Collision of Dna, Race, and History. *Rutgers University Press*. pp. 32–40.

Leeds, J. P. 1991. Moore v. Regents of the University of California: More for Biotechnology, Less for Patients. *Indiana Law Review*, 25(2), pp. 559–588.

Leonelli, S., 2010. The commodification of knowledge exchange: Governing the circulation of biological data. in Radder, H. (Ed.) *The commodification of academic research: Science and the modern university*. University of Pittsburgh Press. pp. 132–157.

Locke, John. 1960 [1698]. *Two Treatises on Government*. Cambridge: Cambridge University Press.

M'charek, A. 2021. Traces of Race, Roots of Gender. *The Routledge Companion to Black Women's Cultural Histories*. London: Routledge.

Mirowski, P. 2011. *Science-mart: Privatizing American Science*. Cambridge: Harvard University Press.

Mitchell, R. and Waldby, C. 2006. *Tissue Economies: Blood, Organs, and Cell lines in Late Capitalism*. Durham: Duke University Press.

Mitchell, R. and Waldby, C. 2010. National biobanks: clinical labor, risk production, and the creation of biovalue. *Science, Technology, & Human Values*. 35(3), pp. 330–355.

Murphy, H. 2018. What 13,000 Patents Involving the DNA of Sea Life Tell Us About the Future. *The New York Times*, September 17. https://www.nytimes.com/2018/09/17/science/patents-marine-dna.html.

Nelkin D, Lindee S. 1995. *The DNA Mystique: The Gene as a Cultural Icon*. New York: Freeman

Niewöhner, J. 2011. Epigenetics: Embedded Bodies and the Molecularisation of Biography and Milieu. *BioSocieties*, 6(3), pp. 279–298.

Ong, A. 2010 'An Analytics of Biotechnology and Ethics at Multiple Scales', in Ong and Nancy N. Chen, eds, *Asian Biotech: Ethics and Communities of Fate* (Durham NC: Duke University Press, 2010).

Ong, A. 2016. *Fungible life: Experiment in the Asian city of life*. Duke University Press.

Parry B. 2002. Cultures of Knowledge: Investigating Intellectual Property Rights and Relations in the Pacific. *Antipode* 34(4):679–706

Parry, Bronwyn. 2004. *Trading the Genome: Investigating the Commodification of Bio-information*. New York: Columbia University Press.

Parry, Bronwyn and Gere, Cathy. 2006. Contested Bodies: Property Models and the Commodification of Human Biological Artefacts. *Science as Culture*, 15(2), pp. 139–158.

Parry, Bronwyn and Greenhough, Beth. 2017. *Bioinformation*. New York: Wiley.

Parthasarathy, S. 2017. *Patent Politics: Life Forms, Markets, and the Public Interest in the United States and Europe*. University of Chicago Press.
Pavone, V. and Goven, J., eds. 2017. *Bioeconomies*. Cham: Palgrave Macmillan.
Pollock, A. 2019. *Synthesizing Hope: Matter, Knowledge, and Place in South African Drug Discovery*. University of Chicago Press.
Pottage, A. 2004. Introduction: The Fabrication of Persons and Things. In Pottage, A. (ed.), *Law, Anthropology, and the Constitution of the Social: Making Persons and Things*. Cambridge: Cambridge University Press.
Pottage, A. 2007. The Socio-legal Implications of the New Biotechnologies. *Annual Review of Law and Social Science*, 3, pp. 321–344.
Quigley, M. 2018. *Self-ownership, Property Rights, and the Human Body: a Legal and Philosophical Analysis (Vol. 43)*. Cambridge University Press.
Rabinow, P. 1999. Artificiality and Enlightenment: From Sociobiology to Biosociality. In *Essays on the Anthropology of Reason*. Princeton: Princeton University Press.
Rabinow, P. 2011. *Making PCR: A story of biotechnology*. University of Chicago Press.
Rabinow, P. and Rose, N., 2006. Biopower Today. *BioSocieties*, 1(2), pp. 195–217.
Radin, J. 2018. Ethics in Human Biology: a Historical Perspective on Present Challenges. *Annual Review of Anthropology*, 47, pp. 263–278.
Rajan, K. S. 2006. *Biocapital: The Constitution of Postgenomic Life*. Durham: Duke University Press.
Rajan, K.S. 2017. *Pharmocracy: Value, politics, and knowledge in global biomedicine*. Duke University Press.
Raman, Sujatha and Tutton, Richard. 2010. Life, Science, and Biopower. *Science, Technology, & Human Values*, 35(5), pp. 711–734.
Reardon, J. 2009. *Race to the Finish: Identity and Governance in an Age of Genomics*. Princeton University Press.
Reardon, Jenny and TallBear, Kim. 2012. "Your DNA Is Our History" Genomics, Anthropology, and the Construction of Whiteness as Property. *Current Anthropology*, 53(S5), S233–S245.
Rose, N. 2001. The Politics of Life Itself. *Theory, Culture & Society*, 18(6), pp. 1–30.
Roy, V. 2020. A Crisis for Cures? Tracing Assetization and Value in Biomedical Innovation. *Assetization: Turning Things into Assets in Technoscientific Capitalism*, pp. 89–116. New York: MIT Press.

Schwartz-Marín, Ernesto and Méndez, Alberto Arellano. 2012. The Law of Genomic Sovereignty and the Protection of Mexican Genetic Patrimony. *Medicine and Law*, 31, p. 283.
Sell, S. 2003. *Private Power, Public Law: The Globalization of Intellectual Property Rights*. Cambridge: Cambridge University Press.
Sherman, B. 2008. Taxonomic Property. *The Cambridge Law Journal*, 67(3), pp. 560–584.
Simpson, B. 2013. Managing Potential in Assisted Reproductive Technologies: Reflections on Gifts, Kinship, and the Process of Vernacularization. *Current Anthropology*, 54(S7), pp. S87–S96.
Slabbert, M. Nothling and Pepper, Michael S. 2010. A Room of Our Own: Legal Lacunae Regarding Genomic Sovereignty in South Africa. *THRHR*, 73, p. 432.
Strathern M. 2001. The Patent and the Malanggan. *Theory Cult. Soc.* 18(4):1–26
Strathern, M. 2005. *Kinship, Law and the Unexpected: Relatives Are Always a Surprise*. Cambridge: Cambridge University Press.
Taussig, K.S., Hoeyer, K. and Helmreich, S., 2013. The anthropology of potentiality in biomedicine: an introduction to supplement 7. *Current Anthropology*, 54(S7), pp. S3–S14.
Winickoff, D. E. 2015. 2 Biology Denatured: The Public-Lives of Lively Things. In Hilgartner, S., Miller, C. A., and Hagendijk, R. (eds.), *Science and Democracy: Making Knowledge and Making Power in the Biosciences and Beyond*, p. 15. London: Routledge.

5

Brains: Self and Personhood

> The idea that 'we are our brains' is not a corollary of neuroscientific advances, but a prerequisite of neuroscientific investigation.
>
> Vidal, "Brainhood, Anthropological Figure of Modernity"

> Neuroscience is a purely mechanistic science. Neurons, neural networks, and the connectome do not have reasons. They have no aspirations, no sense of past, present, and future. These are properties of agents. Ethics and law are addressed to agents. Thus, there will always be a problem of translation between the pure mechanism of neuroscience and the folk psychology of ethics and law.
>
> Morse, "Neuroethics: Neurolaw"

Introduction

For some scholars, the impact of neuroscience on law has the same promise as genetics did in the 1980s (Farahany 2015; Garland and Frankel 2006). Like the genetic "truth machine," neuroscience promises to render the brain "transparent": Once we "scan" your brain we can

S. van Wichelen and M. de Leeuw, *Biolegality*, Biolegalities,
https://doi.org/10.1007/978-981-99-8749-8_4

see "who" you really are! In contemporary public discourse, the prefix "neuro" seems to be added to every kind of self-help manual, whether we are talking about neuro-management or brain-based parenting (O'Conner et al. 2012; Mackenzie and Roberts 2017). Moreover, neuroplasticity seems a welcome addition to the neoliberal management of everyday life, revealing how "adaptability"—our internal change management—now reaches beyond the social into the biological; if we just work on ourselves, we all can have the brain we want![1]

The relation between neuroscience and law seems to reflect an ongoing "naturalism" in legal thinking. As described in Chapter 2, this naturalism derives from scientific realism, which believes science reveals the physical or evolutionary reality of human behavior based on empirically verifiable facts. Contemporary neuroscience promises a new level of insights into the systemic neurological, behavioral, and psychopathological preconditions of human conduct. Approached through a biopolitical framework, neuroscience governs another layer of life, bioethics examines the normative contours of neuroscientific research, and neurolaw focuses on the legal implications of the neuroscientific mind/brain debates. Biolegality offers a lens for understanding law and neuroscientific knowledge as mutually determining each other's validity: where law validates the outcome of neuroscience, neuroscience validates legal categories.

Using neuroscience in law is not new (Bigenwald and Chambon 2019; Morse 2006, 2011b; Sapolsky 2004). Since the early twentieth century, medical-legal disputes about the definition of "insanity" rest on the differentiation between *mens rae*, a guilty mind, from an *actus reus*, a guilty act. The 1960s saw the rise of the "neuromolecular" gaze in public consciousness, as the spread of neuroscientific findings signified a shift from the psychological to the neurobiological self (Rose and Abi-Rached 2013). This shift was further solidified by the litigation lawyer J. Sherrod Taylor who in 1995 coined the term "neurolaw." Although in the 1990s the use of MRI (Magnetic Resonance Imaging) machines was still rare, the use of neuropsychological expertise to explain "brain structure, brain function, and how the alleged tortious injury was likely related to abnormal brain function" rapidly increased in civil brain-injury litigation. Taylor noticed that bringing neuropsychological expertise into the courtroom could change "a 'sprain and strain' case (settling for less than $10,000) into a brain injury case (with a settlement in the hundreds of thousands of dollars)" (see p. 687 in Shen 2016). Besides starting "The Neurolaw Newsletter" in 1991, Taylor's article "Neurolaw: Towards a New Medical Jurisprudence" (1995) and his book *Neurolaw: Brain and Spinal Cord Injuries* (1997) firmly established the new field and became a central resource for showing lawyers how best to use neuro-medical science and jargon in the courtroom. As a result, neurolaw increasingly found a stronghold in the US criminal justice system in the 2000s (Garland 2004).

Neuroscience's integration in criminal proceedings now covers a wide range of problems and fields claiming to explain human behavior; from lack of impulse control, sexual aberration, predicting the chance of recidivism, detecting insincerity, culpability awareness, to bioenhancement of our moral attitudes. Scholars note that neuroscience biologizes socialization and responsibility, limiting the impact of upbringing, education, social environment, gender, and so on (Slaby and Choudhury 2018). A growing focus in neurolaw is on the examination of brain maturity: how can neuroscientific knowledge of brain development be linked to legal standards of intentionality and culpability to establish—beyond reasonable doubt—a suspect's guilt or innocence? To test this question juvenile suspects are placed in an MRI machine providing visual data with the promise of more objective and reliable scientific knowledge than the subjective diagnosis of a psychologist (Federle and Skendelas 2009; Sánchez-Torres et al. 2024). Instead of psychological therapy or social re-integration the suspect or prisoner is now offered a neuro-diagnosis and a biomedical intervention and treatment plan. Like antidepressants, anti-aggression medication proposes a technological fix, enabling us to again function among other humans while treating "aggression" as a mostly neuro-medical problem.

Do novel developments in neuroscience reveal new explanations for human behavior and does it redefine human agency, legal personhood, and responsibility? Or does neuroscience, instead, contribute to an intensified biological reductionism that decontextualizes and depoliticizes identity and agency? While this chapter starts by describing the deeper philosophical and scientific disputes regarding the metaphysics of free will, it ultimately aims to exit this debate to more productively analyze how neuroscience impacts law and society. As we will see, positions on free will are mostly determined by philosophical or normative "camps" rather than scientific evidence, although it is the appeal to the hard sciences that gives determinism its strength. Absolute biological determinism claims that our behavior is bound by the bio-physical laws governing life on earth, which means that our "mind is dependent, in a crucial way, on its biological basis" (see p. 18 in Pernu and Elzein 2020). From this perspective, the ultimate question is whether this biological basis is flexible or static, whether it allows for forms of self-control and self-awareness.

Instead of extending the debate on free will, this chapter focuses on the empirical ways law and neuroscience are now entangled (or not), and how this entanglement impacts our social understanding of human behavior and criminal conduct. Starting from a recent appeal in Texas to halt the execution of a man convicted of a crime he committed at 18, we examine the ways neuroscience becomes legitimate evidence in criminal courts. From a biolegal perspective, we identify how two different regimes of *validation*—biological and legal—create tensions in how we conceptualize agency, responsibility, maturity, and human morality (among others). Like the "truth machine" of forensic genetics, neuro "machines" (such as lie detectors and fMRI scans) appear to offer a similar expectation of objectivity, but at this point do not (yet) carry the same weight as genetic testing. Whereas a *neurofication* is taking place in society—and law is implicated as an important institution that is both impacted by and producing neurofication—we describe how instead of producing stable "neuro-truths," neuroscientific practices produce neuro-legal categories that continue to shift rather than become fixed. We also take up the question of posthuman enhancement figures in relation to law and neuroscience, highlighting some of the problems associated with the notion of "brainhood" (the idea that our selfhood is entirely determined by our brain). Rather than advocating brainhood, we discuss the concept of neuro-rights and show how a relational framework around neuroscientific knowledge could sustain a more critical approach to incorporating neuroscience in law.

Neurofication in Law

At 18 years old Billy Joe Wardlow attempted to leave Texas with his girlfriend to flee from their abusive homes. While trying to steal a truck in rural Morris County, he got into a struggle with Carl Cole, the truck's owner. During the fight Wardlow shot Cole. Lincoln Caplan summarizes the case aptly, in *The New Yorker*:

> At his trial, in 1995, Wardlow testified that he had intended to intimidate Cole, not kill him, when he brandished a .45-calibre pistol he had stolen from his mother. Cole grabbed Wardlow's arm, and Wardlow fired the gun, shooting Cole between the eyes. A jury convicted Wardlow of capital murder and sentenced him to death. Until then, he had never committed a violent offense. To sentence him to death in Texas, the jury had to decide that he "would commit criminal acts of violence that would constitute a continuing threat to society:" in other words, the jury had to belief 'beyond reasonable doubt' that Wardlow remained dangerous in the future.[2]

Wardlow spent 25 years on death-row before being executed on July 8, 2020.[3] A few weeks before the execution several Texas lawmakers, supported by a group of prominent neuroscientists and psychiatrists, filed a petition (*Billy Joe Wardlow vs State of Texas*) to overturn the capital punishment verdict.[4] They objected to the legal notion of "future dangerousness" used to justify the death penalty. The lead neuroscientist made two interlinked arguments: first, new neuroscientific evidence shows that developmentally the brain of an 18-year-old—the legal age of maturity and full culpability—does not differ from the brain of a 17-year-old. Due to Wardlow's actual "brain immaturity" he should not have been treated as a legal adult:

> [New] neuroimaging studies reveal that the prefrontal cortex—an area of the brain associated with reasoning and executive function—remains developmentally immature and underregulated until the mid-20s, while the brain's reward centers are relatively overexpressed, making emerging adults 'more vulnerable to impulsivity', less capable of emotional reasoning, and more likely to make 'errors in self-regulation'. Because of

> these developmental delays, the judgment and decision-making of adolescents and emerging adults differ from adults' in several respects: they are less able to control their impulses; they weigh the risks and rewards of possible conduct differently; and they are less able to envision the future and apprehend the consequences of their actions. (see p. 13 in *Wardlow vs Texas*)

Their second argument logically follows the first: because the brain of an 18-year-old cannot be perceived as "mature" and settled, it is not possible to predict any "future dangerousness" or establish "permanent incorrigibility" from it: "It is impossible for experts to distinguish whether an emerging adult who engages in risky or impulsive behavior is merely evincing a neurological immaturity or, rather, exhibiting signs of a dangerous future" (see p. 14 in *Wardlow vs Texas*).

The Texas Supreme Court rejected the petition, arguing there was no proof of a constitutional or procedural problem with the original verdict, hence, no reason to halt the execution. If Wardlow had been 17 years of age at the time he committed the murder, the 2005 US Supreme Court verdict in *Roper vs Simmons*[5] could apply. *Roper vs Simmons* made it illegal to impose the death penalty for crimes committed under the age of 18 because it violated the 8th Amendment against *cruel and unusual punishment*. The US Supreme Court's narrow 5–4 verdict focused on the so-called "evolving standards of decency" test which, the majority opinion stated, shows that—based on sociological and scientific research—juveniles lack maturity and a sense of responsibility. Wardlow being 18 at the time of the crime allowed the court to declare there was no constitutional need to halt the execution (this was a simple legal rebuttal ignoring the neuroscience and psychiatric experts' argument concerning the similar brain immaturity of 17- and 18-year-old males). But more troublesome as the legalistic rejection was the ongoing acceptance of Wardlow's future "dangerousness" based on a psychiatric assessment now perceived as "junk science."

This tension between neuroscience and law exposes a struggle between knowledge *validation* and legal *justification*. Whose normative truths should prevail? How do scientific notions of brain maturity engage legal criteria of criminological culpability? How does neuroscience determine the standard for brain maturity? Do legal and neuroscientific concepts of risky behavior and "future dangerousness" correlate or is their "correlation" a social construction to match two entirely different methods (science vs legal-moral) with each other? Biolegality considers many legal concepts as "factional": fictive constructs with factual, material consequences. While neuroscience considers its research (the brain of an offender) as revealing a material truth, the question of legality remains a question of interpretation and application of collectively agreed upon, constructed legal procedures.

The Wardlow case reflects a growing concern within neurolaw around the question of brain "maturity" and individual responsibility.[6] Nonetheless, at present, neuroscience does not have hard standards for defining maturity in either normative (e.g. a specific age) or empirical terms, much less for predicting "future dangerousness." Instead, neuroscience can only deliver "data" by way of graphs, scans, chemical reactions, etc. This data does not say anything about dangerousness or social behavior; it merely stands in a certain relation to other data, especially a base line that represents a "normal" brain condition. But as Stephan Morse aptly summarizes, law's persons are rational or irrational regardless of the brokenness of their brains: "If the defendant's brain appears broken, but he is a rational agent, he is rational for legal purposes. If the brain appears normal, but the agent is clearly psychotic, the agent is not rational for legal purpose" (see p. 339 in Morse 2016). Using neuroscience in law—much like the case discussed here—has been regarded as reductive in two ways. First, in a positive sense, neurobiological arguments claim to overcome law's subjective and potential arbitrary criteria by revealing biological "facts." On the other hand, and in a more negative vein, neurological "facts" may ignore the social, personal, and historical context of events, thereby devaluing human agency, autonomy, emotions, and moral selfhood. Nikolas Rose offers a typical biopolitical perspective on reductive reasoning whereby the legal trumps the biological due to its overriding legal-political necessities:

> When the judiciary defends the non-genetic, non-psychiatric fictions of free will, autonomy of choice and personal responsibility, this is not because legal reasoning considers this to be a scientific account of the

> determinants of human conduct. Rather, the criminal justice system deems it necessary to act on the basis of this image of the human being for reasons to do with prevailing notions of moral and political order. Indeed, the trend of contemporary legal thought, especially in the USA, is to operate on the premise of the *inescapability* of moral responsibility and culpability. (Rose 2010: 84, emphasis ours)

It seems the Texas Supreme Court took exactly this position: *whatever* the science states, Wardlow's accountability had to remain "inescapable." Still, the reverse might evoke equally troubling consequences: if neuroscience shows an offender's brain is unable to change due to the settled ("mature") neuropathological structures or irreversible traumatic impact, the convicted person is condemned to an unchangeable "nature" by scientific proof. Here, any "therapeutic" approach goes both ways: triggering voluntary or involuntary medical intervention or a call to never release the incurable "patient." The "inescapability" of a sentence for certain groups echoes ongoing racist and past eugenic practices dehumanizing and criminalizing people independent of their actual acts.

Looking at US criminal case law, Farahany (2015) shows that between 2005 and 2012 the incidence of brain science in judicial reports more than doubled. Indeed, there is ample empirical evidence to suggest that the use of neuroscience in criminal defense is growing and that more and more defendants turn to science to argue that their neurological condition affected their moral responsibility. A more recent study by Greely and Farahany shows that only a small fraction (20%) of all judicial reports from 2015 included neuroscientific arguments in cases involving capital punishment, whereas 40% involved cases other than homicide (see p. 455 in Greely and Farahany 2019). Although only 10% use direct brain scans in their defense and 40% did not resort to direct testing, the claim that "the brain did it" dominated these reports (Farahany 2016). Nonetheless, the growth of neuroscientific claims does not necessarily imply that bioscientific arguments are superior. As Rose stipulates, "defence lawyers [...] utilize anything they can to defend their clients" (see p. 13 in 2000). The authority of science in the public domain provides lawyers with a strategic avenue to tap into the epistemic power that is usually reserved for experts.

At the same time, the influence of neuroscience on actual legal outcomes is much smaller than its current debate in criminology and bioethics assumes (Morse 2017). While assessing mental health or brain injuries is crucial in many criminal cases (to establish and evaluate mitigating circumstances), these are usually diagnosed by a forensic psychologist or psychiatrist. Neuroscientists themselves rarely appear in court rooms. It is appropriate, considering this gap, to establish where the desire to scientifically objectify human behavior comes from. While free will assumes we act voluntarily and, hence, are responsible for what we do, the social and political context impacting our agency remains difficult to objectify or quantify (what percentage of my aggression was caused by my violent family upbringing?). Neuroscience promises a more objective measure to determine the reason for criminal behavior. More progressively, we can view the push to neuroscientific frameworks as advocating for a more therapeutic jurisprudence replacing incarceration with psychiatric treatment. Neuroscience can also help explain crime through a social and environmental lens: if, for example, an alcoholic parent or chemical toxins under the house impact my neurological condition, this "neuro-knowledge" can explain my mental state.

Regardless of the explanatory bases that inform the field, the current legal scholarship on neurolaw points to a neuroscientific reductionism that revises the foundation of responsibility by offering a "neurologised" version of the legal person replacing both the free liberal individual of rights and the socio-economic determined person with a now primary "biological makeup" (see p. 46 in Bigenwald and Chambon 2019). Does being a violent or empathetic person merely depend on the presence or absence of certain neurological pathways (having a "normal" frontal cortex)? Ultimately, deliberations like these lead to the question of whether neurolaw triggers biological reductionism as pure materialism (the mind is the brain) such that we can therefore speak of guilty brains instead of guilty persons (Hodgson 2000).[7]

Guilty Brain or Guilty Person: The Boundaries of Responsibility

To establish legal culpability is both a complex and ambivalent task often simplified by the difference between the guilty mind and the guilty act:

> For you to be held legally responsible, a harmful event must have occurred, and that event must have resulted from actions that you wilfully and freely decided to perform. That is, the right sort of causal connection must hold between your decisions to perform certain actions and the outcomes of those actions. (see p. 3 in Pernu and Elzein 2020)

From a historical perspective, Pernu and Elzein show that holding a person accountable for their actions is a modern phenomenon—the *intention* of the individual was never at issue in the premodern world (see p. 4 in 2020, and Vincent 2013). In modern criminal cases, the situation is reversed, and the focus falls primarily on "assessing the level of culpability of the defendant—even to the point of ignoring the issue of whether a harmful outcome actually resulted from the defendants actions" (see p. 4 in Pernu and Elzein 2020). Although there is a clear distinction between what we mean by "agency" and "free will" it can be hard to mark the exact point of differentiation. We all act and respond, but do we know exactly why we act the way we do? In law this differentiation is decisive to establish levels of innocence and guilt. If a suspect hallucinates while hitting another person, the guilty act lacks guilty intention; if a psychotic person hears voices commanding them to harm another person, punishment is mitigated. The recognition of these forms of cognitive impairment is crucial to avoid imprisoning persons who need mental health treatment instead.

Moreover, many in the field of neurolaw seem to adhere to a form of "ontological nomism," i.e. the idea that (neuro)science can reveal the metaphysical truth of who we are and why we act in certain ways:

> According to a naturalistic, scientific, world view, reality is ultimately physical. Therefore, the human mind—our decision-making and behavior—must also be fundamentally physical. It would seem to follow,

> then, that the neurosciences, that study the physical basis of our minds, should be directly useful in understanding human decision-making and behavior, and should therefore also inform our moral and legal judgements. (see p. 1 in Pernu and Elzein 2020)

This idea assumes that once we are aware of and understand neurological conditions we can explain, for instance, a suspect's lack of impulse control, allowing the court to conclude that the brain condition, not the person's free will, is to "blame" for causing harm. The tension between a general theory (a less developed frontal cortex can result in less impulse control), and individual differences (not all people with a less developed frontal cortex lack impulse control) mirrors the same tension in law, which applies general principles (all murderers go to prison) to individual cases (*this* murderer is less culpable due to mental disability). This can lead to a general scientific "truth" being used as a general legal "principle": all persons X with a Z-size frontal cortex will endanger society due to high risk of reoffending (because *generally* unable to control their impulses). It can also go the other way: due to a specific neural condition this person must not be treated as a criminal, but as a person in need of medical help. It is the boundary work distinguishing between a potentially dangerous person or a person with a mental health problem that is crucial here, as well as how legal principles ("equality before the law") and legal form ("future dangerousness") are applied to the defendant.

Taken at its extreme, a pure physicalist position would argue that all human behavior is neurological and therefore all actions can be excused: it was never "me" but always my brain "who did it" (see pp. 13–15 in Pernu and Elzein 2020). Or, in reverse, due to "me" being my 'psychopath' brain "I" am always guilty. There are ample cases in the past and present that show that dramatic forms of brain injury cause radical personality changes (see p. 14 in Pernu and Elzein 2020).[8] Pernu and Elzein caution, however, that such cases assume that biological basis is the sole cause for radical behavioral variations. Thus, they instead contend that a physicalist approach is not sufficient to understand the mind: "you also need to be in possession of the complete description of the body's surroundings, and the history of the interactions of the two" (see p. 17 in ibid.). While the debate between free will and determinism continues between psychologists, philosophers, and ethicists, in practice, boundaries are being drawn regardless of the position one takes in this debate. Technology plays an important role in drawing these boundaries.

Machines of the Brain: Science and Sincerity

Neuroscientific knowledge mostly results from measurements through imaging technologies, such as computed tomography (CT) scans, positron emission tomography (PET) scans, electroencephalograms (EEGs), functional magnetic resonance imaging (fMRIs), and X-rays. Used increasingly in courts, these technologies examine the size of the prefrontal cortex and brain tumors, the impact of epileptic seizures, drug or alcohol addiction, and side effects of medication, all of which inform how law and brain biology mark and validate certain human behavior. Courts then adjudicate whether the displayed behaviors were intentional or not (Pitts-Taylor 2016; Rees 2016; Rose and Abi-Rached 2013). Such screening technologies might travel the same route as DNA and become a legally admissible form of evidence (Dumit 2004). Wrapped up in these processes and practices are ideas about scientific standards, measurements, statistical analysis, and the definition of scientific as well as legal evidence. The legal brain, then, becomes a prime field in which legal and biological knowledge simultaneously challenge and produce each other's validity (Heinemann and Heinemann 2010; Pickersgill 2013) through forms of technological mediation. The increasing use of brain scans promises both retrospective and predictive evidence to explain why a person committed certain acts (and might repeat them if the brain is seen as innate and static).

One specific technology advanced by neuroscience pertains to evidence of deception. Determining if a suspect, witness, or victim is deceptive (lying) or sincere (truthful) goes to the core of all judicial systems and the current surge in the use of EEGs in courts illustrates a desire to find new, more objective ways to reveal deception by suspects, victims, or witnesses. In doing as if the brain can ultimately be made transparent, "the truth" is forced out of the person. For centuries, torture was the preferred violent method to elicit truth and disprove "lies" (whereby "torture truths" are usually lies due to giving the torturers what they want to make them stop). The polygraph machine (colloquially called the "lie detector") took over the test of sincerity. A polygraph registers the connected person's pulse, blood pressure, and respiration assuming that by answering a set of questions "lying" becomes a detectable, physiological measurable, condition. But physiological "lie detection" is mostly considered a pseudo-science.[9] This does not, however, prevent US security agencies from using the technology to screen employees (see p. 208 in Kahn 2015).

While polygraph and EEG testing measure *external* responses (represented by waves),[10] the introduction of functional Magnetic Resonance Imaging (fMRI) allows a look *inside* the skull to "see" the brain as its neural pathways respond to external inputs. Here, the subject lies in the hollow space of the MRI machine and is presented with images and asked to respond to questions, which in turn trigger certain brain fields to activate. This activation manifests as a visual representation ready for interpretation. Kahn summarizes the process as an "ordeal by radio waves":

> The MRI uses a pulse of radio waves to temporarily knock protons out of alignment in a targeted area and then uses the magnet to measure the radio waves they emit as they relax back into alignment. Computers then take this data and process it to create an image of the subject's brain. (see pp. 208–209 in Kahn 2015)

Nonetheless, the use of fMRI scanning runs into two main critiques. First, the test subjects are mostly WEIRD (Western, Educated, Industrialized, Rich, and Democratic) students at Western universities, hardly a representative segment of the global human population. Moreover, fMRI outcomes do not directly represent the brain as such, but are, "highly complex, technologically mediated statistical constructions of results that have many steps involving human interventions to set parameters and construct results in a manner that can actually be highly subjective" (see p. 210 in Kahn 2015). Studies have questioned the correlation between certain brain activities and forms of human behavior, indicating a strong preference for results that fit best with the goals of the research design (see

p. 211 in Kahn 2015; Margulies 2012; Dumit 2012). In fact, neuroscientists themselves acknowledge that images are actively manipulated to adequately display the data they are researching (Dumit 2012). Beyond these issues of technical mediation and selection, according to Dumit, the effect of seeing brain images contribute to understanding relations of personhood and agency as measurable and visualizable (2012: 119, see also Alac 2008).

Another problematic use of fMRI aims to detect conscientious objection or the "sincerity" of one's beliefs and convictions. Examining the landmark *Burwell v. Hobby Lobby* case (2014),[11] Kahn describes how fMRI scanning was proposed for testing "the sincerity of the experience of the burden" in future cases (see pp. 217–220 in 2015). In this case, the US Supreme Court ruled that private corporations are exempt from certain employment regulations on religious grounds. The Hobby Lobby case concerned the requirement that the craft chain store provide employee health insurance plans that meet government mandated standards, which included reproductive care for female employees. The proposed use of fMRIs would supposedly enable courts to determine the sincerity of the claimed "burden." Such use, as Kahn rightly argues, attempts to make "inner life" (religious belief, sincerity, conscientious objections, honesty) neurobiologically detectable.

Despite claims to render objective evidence of inner beliefs, brain images also have a cultural life outside of the courtroom. Especially in common law countries, juries bring cultural understandings of technical images to courts. As Dumit puts it: "The jury's eyes are cultural ones, not expert ones" (see p. 132 in Dumit 2004). Such cultural conceptions are especially pertinent when we recognize that the main methodological assumption guiding brain machines is *correlation.* Correlation does not equal causation and any claims grounded in "facts" generated by correlation require careful analysis. In addition, once we acknowledge the clear gap between legal and scientific "facts," the difference between correlation and causality is central to assessing future uses and abuses of neuro-technologies and their impact on law.

The Posthuman Problem of Enhancement

One of the future applications of neuroscience involves brain-based self-enhancement therapies. Neuro-enhancement describes a wide range of "tools," from our morning coffee to strong opiates suppressing pain, from antidepressant to party pills, and from anti-anxiety to study focus aids. Some self-administered neuro-enhancement "pills" are widely accepted, others, often with similar substances in them, illegal. Like antidepressants that aim to battle mental health conditions by adjusting serotonin levels, it is possible that legally prescribed anti-aggression medication to control our impulses or to cultivate empathy toward others might become available soon (Butler et al. 2021; Settumba et al. 2021). Neurolaw literature explores the potential for state mandated neuro-enhancement as a legal alternative to punishment. While society is rife with substances for enhancing the body and cognition, these enhancements are voluntary and not enforced by the court.

In the criminal justice system "enhancement" refers in particular to the potential use of *moral* bioenhancement through, for instance, the forced use of psychotropic drugs, brain stimulations, neurosurgery, or gene editing with the intent to lower aggression or improve empathy. The ethical and philosophical quandary revolves around whether the law should coerce people into taking medication that would change their "moral biology." Bioethicists Baccarini and Malatesti (2017) argue that mandatory moral neuro-intervention on psychopaths is logically justifiable since, if given the chance, psychopaths themselves would prescribe coercive drugs to others. Reversing the Kantian maxim to *not do to others what you don't want done to you*, these authors argue that what the psychopath would do to others can also be done to them. Others notably disagree, arguing that the narcissistic nature of psychopaths contradicts the maxim and, hence, mandatory moral bioenhancement of psychopaths signifies a form of violence and can ethically not be justified (Sirgiovanni and Garasic 2020). Only the utilitarian argument of the "greater good" seems morally justified: can psychopaths be "morally enhanced" to become "better people" through medicine to keep society safe? While "betterment" is usually seen as resulting from access to education, therapy, and social support, biologized interventions in the material condition of our brains are becoming increasingly possible.

According to Kudlek, neuroscientific research has produced findings supporting the potential correction of immoral behavior:

> These findings range from the administration of neurohormones such as oxytocin (which promotes pro-social attitudes like trust, sympathy, and generosity), to the manipulation of serotonin and testosterone levels (which can mitigate undue aggression, and increase fair-mindedness, willingness to cooperate, and the aversion to harming others). Brain stimulation devices have also proven useful in morally relevant areas by reducing impulsive tendencies in psychopaths, treating addiction, and improving self-control. (see p. 102 in Kudlek 2022)

Besides the legal-psychological questions of whether and when mandatory treatment is allowed and what authority determines who is a psychopath, the question of what moral standards are used to "enhance" a person remains. To state it bluntly, many people could use a bit of moral enhancement, but one fundamental presupposition of Western liberal societies is that we do not round people up to subject them to medical interventions to "better" them.[12] Having said this, we must acknowledge relatively recent medical interventions, often racialized or focused on disabled or "deviant" groups like homosexuals, migrants, or Indigenous populations intended to morally, sexually, and genetically "enhance" these populations to a "white" Western cultural heteronormative standard. Medical experimentation on black males like the Syphilis Study at Tuskegee[13] still occurred until the early 1970s despite the total disqualification of eugenic research and politics after the Holocaust. But the lack of medical intervention, for example, by excluding non-white groups or populations in the Global South from medical research or pharmaceutical trials also exemplifies the relation between biological and legal "standing."

A broader approach to neuro-enhancement emerges from the fields of biosocial criminology and neuro-criminology. Here the focus is on preventing violence or reoffending through a better understanding of the relation between neurological and mental attributes and criminal violence. Criminology combines, in this instance, with neuroscience, psychology, and public health. While the social and environmental aspects of community violence have been studied at length, neuroscientifically inclined criminology scholars argue that the neurological and mental health background of juvenile offenders is understudied. Neuro-criminology intends to connect early childhood abuse, juvenile mental health, and adult offending, showing how they might result from brain trauma or early neurological deficits like ADHD or lack of impulse control. While social and educational support is still seen as primary, understanding and diagnosing the neurological aspects of potential violent behavior justifies medical "corrections" by way of pharmaceutical (preventive) interventions (Spenser et al. 2019; Chang et al. 2016; Lichtenstein et al. 2012).

From Brainhood to Neuro-rights

If "we *are* our brains" all forms of personhood, self, and agency become legible, first and foremost, through diagnostic descriptions of our neurological condition, producing an image of our "neuro-existence," our neuro-selves. An obsession with the brain, as the organ representing the location of our mind, reasoning, rationality, reflexivity, and consciousness goes back to at least the seventeenth century when David Hume, John Locke, and René Descartes replaced the soul with the brain/mind, religion with science, and moral speculation with causal and logical analysis. As a result, the body gets abstracted from the mind, and the "person" transfers to a medical-biological system subject to, often absurd, philosophical mind-puzzles (if we put the brain of a prince in a cobbler does the cobbler become a prince, or not?). Since Hume's skeptical empiricism and Descartes' *ergo cogito sum*, philosophical speculations concerning human consciousness and agency were slowly taken over by neuroscientist cutting and slicing of the material brain. For Francisco Vidal, brainhood "predated reliable neuroscientific discoveries, and constituted a motivating factor of the research that, in turn, legitimized it" (see

p. 14 in Vidal 2009). Hence, the history of the brain sciences is intertwined with the rise of the cerebral subject. While Vidal observes that the phenomenon of brainhood is historically exclusively Western, it is now firmly globalized and hegemonic (see p. 11 in Vidal 2009).

More recently, STS and feminist scholarship have emphasized neuroscientific research into brain plasticity and the brain's entanglement with the gut (Pitts-Taylor 2016; Rees 2016; Wilson 2015), challenging the reductionist account by demonstrating the role environmental factors play informing new and/or alternative neurological pathways. More broadly, social science scholars call for a "critical neuroscience" situating brains and bodies in a broader bio-psycho-social context, one that underscores the brain's relation to other parts of the body and to economic, cultural, and psychosocial environments (Slaby and Choudhury 2018).

Finally, while enhancement might be considered a potential intervention in mental or even public health, enforcing a socio-neurological norm in the body itself clashes with the right to bodily autonomy beyond legally established punishments (like being put in prison for harming others). This triggers the question if and how our rights must be extended to cover "neuro-rights," namely the right to control the makeup of one's brain or neurological condition as key part of personhood. Involuntary intervention in the brain or body echoes the historical injustices of biological interventions such as (chemical) castration of homosexuals. In fact, until recently, California and Florida still permitted the "chemical castration" of sex offenders to quell their sex drive (Mascolo and Bui 2023) and forced sterilization of mentally disabled women was still taking place in the US and Sweden (Reilly 2015)—practices that clearly undermine personhood rights and human dignity. Neurotechnology's potential to directly intervene in our brains, download brain data from implants, or send data back into the brain manipulating our memory, speech, and thinking, is no longer science fiction, but actively being tested in private neuro-labs in Silicon Valley. Even if neurotechnology is used to restore neural pathways of Parkinson's patients the brain data this delivers is, ultimately, just that: data waiting to be commodified and hacked. This triggers the urgent need for legal frameworks regarding brain privacy, protecting of our sense of self, or the right of equal access to neuro-treatment or health insurance. The idea that our brains can be "hacked" might sound surreal but is part of daily experiments on rats in neuro-labs around the world, the knowledge that we are next triggers urgent legal questions regarding our neuro-rights in the age of brainhood.

Conclusion

In sum, we can say, more broadly, that connecting neuroscience and law creates a new "sociotechnical imaginary" in which "neurolegal discourse" produces "new visions of law, science and scientists" creating an "anticipatory discourse" (see p. 27 in Pickersgill 2011). What this discourse anticipates is the promise of radical objectivity, visualizing—for instance on fMRI scans or EEG lines—how the brain steers and controls behavior and responses. The MRI scan promises to deliver visual "evidence" of emotions, impulses, and intentions, as well as neural signs of empathy, hate, honesty, deception, maturity, trauma, and aggressiveness, clearly and exactly identified by light on scans, ready for expert interpretation. The assurance of objective, scientific diagnosis by way of machines is part of a movement to understand criminal behavior not in socio-cultural, political-economic, and individual contexts but through a biological, neurological fact-finding mission. Insights from science and technology studies demonstrate, however, that machines and images are assemblages, constructions with a social life that traverses multiple sites, institutions, and domains. At present, brain images in neurolaw mostly still figure as "serviceable truth" (Ding 2020; Jasanoff 2015).

Beyond brain tumors, sport or work-related traumas, alcohol or drug impacts, the use of neuroscience data as forensic evidence remains limited. Hence, neurolaw scholars' desire to use neuroscience as a standard tool to discover responsibility, culpability, or mitigating evidence, has not yet penetrated legal institutions and the judiciary. But the growing attention to neuro-ethics, neurolaw, and neuro-rights in courts and criminology clearly indicates that a further bio-legalization of self and personhood in the twenty-first century is afoot. How neuroknowledge and legality intertwine, clash, or create new neuro-legal concepts and practices demands an ongoing critical engagement with the use and abuse of neuroscience in law.

Notes

1. Although less explicitly in public discourse, the inverse is also tacitly communicated: having a certain type of brain makes one destined for poverty or crime.
2. Lincoln Caplan, Should Billy Wardlow be Executed? *The New Yorker*, 30 June 2020, see https://www.newyorker.com/news/news-desk/should-billy-joe-wardlow-be-executed-for-a-crime-committed-when-he-was-eighteen (accessed June 30, 2020).
3. The next day the Texas Tribune headlined the news as "Texas executes Billy Wardlow, who was 18 when he killed a man. Experts argued that's too young for a death sentence" (Texas Tribune, July 8, 2020), which echoed the New Yorker article published a few weeks before.
4. No. 19-8712 *Billy Joe Wardlow vs State of Texas*. Brief of Amici Curiae Professional Organizations, Practitioners, and Academics in the Fields of Neuroscience, Neuropsychology, and Other Related Fields in Support of Petitioner, see https://www.supremecourt.gov/DocketPDF/19/19-8712/145983/20200619160740671_19-8712%20-%20Wardlow%20v.%20Texas%20-%20Professional%20Organizations%20et%20al.%20cert.%20amicus.pdf. The Amici petition represents the final attempt to halt Wardlow's execution after the legal appeals process, which started with a state habeas application in 1998 and continued over two decades with request by Wardlow to abandon further appeals as well as IATC (ineffective assistance of trial counsel) request failed to attain habeas relief. In 2017, three years before his execution, a district court concluded that Wardlow's original counsel was deficient but not prejudiced. In total four courts, state and federal, reviewed and rejected Wardlow's claim before the Amici petition was submitted. For this reason, it was improbable the Amici petition could succeed because it would, besides allowing

novel expert witness and evidence to enter the case at the very last moment, overturn four previous lower-court rulings creating legal inconsistency. In the courts' own language, "prudence counsels against this Court reviewing what four other courts have already deemed wholly without merit." *Billy Joe Wardlow vs State of Texas*, No 19-8835, 21. Wardlow's case shows that the legal process had no real interest in questions of mental development beyond applying existing legal procedures and principles regarding a fair trial or adequate counsel. Wardlow's claim that his "trial counsel was ineffective for failing to conduct any investigation" into his mental health and life story prior to trial, is rejected by the court stating counsel was "reasonable." Although Wardlow had undergone psychological tests by Dr. Walker no finding of mental illness, child abuse or personality disorder were made. Dr. Walker was not called as witness by Wardlow's counsel. This fact was used by the Texas Supreme Court against a stay of execution. The actual core question raised by the Amici petition, that neuroscience shows no difference between brain development of a 17- or 18-year-old, and the impossibility to determine Wardlow's future dangerousness at 18, is not mentioned once in the Court's ruling. Instead, it warrants against further "abusive delay" of the execution. Ibid, p. 39. What Wardlow's case mostly reveals, though, is that neuroscience expertise and legal expertise operate in two different normative registers, and it remains a complex task to align them.
5. *Roper v Simons*, 543 U.S. 551 (2005).
6. Social, biological, and legal standards of maturity change according to time and place, but within the US context, the *Roper vs Simmons* case (2005) introduced neuroscience into the debate.
7. David Hodgson (2000) Guilty Mind of Guilty Brain? Criminal Responsibility in the Age of Neuroscience. *The Australian Law Journal*, Vol. 74, 661–680.
8. Pernu and Elzein recall several historical cases where, due to brain injuries, peoples' character traits changed to such an extent that they were no longer perceived as the same person. In 1848, for example, an iron rod blasted through the left side of Phineas

Gage's face puncturing his forehead. Remarkably the railroad worker survived but after the accident Gage lacked impulse control and became anti-social. In 1966 Charles Whitman shot a person at the University of Texas. Whitman himself did not understand why he did, and requested an autopsy be done after his death. The autopsy revealed a brain tumor that might have caused his sudden urge for aggressive behavior. As Pernu and Elzein rightly conclude, all these historical cases reveal how changes in the prefrontal cortex (PFC) lead to the described changes, confirming PFC as the center of emotional regulation, impulse control, and aggressive behavior. But the reverse seems also possible, as evidenced by a case in which a 30-year-old anti-social and violent male, who in an attempted suicide shot himself with a crossbow through the front of his head, and after surviving, became "docile, social, and 'inappropriately cheerful'" (ibid.).

9. Japan is currently the only country allowing polygraph results in court.
10. The "waves" metaphor refers to the waves the EEG machine makes on the paper, not what actually happens in the brain.
11. *Burwell v. Hobby Lobby Stores, Inc.*, 573 U.S. 682 (2014).
12. One could argue here, however, that mass vaccinations could also be seen as a form of enhancement. This would confirm enhancement as a public health intervention.
13. Between 1932 and 1972 the US Public Health Service (USPHS) conducted the "Tuskegee Study of Untreated Syphilis in the Negro Male" to study untreated syphilis. The study was done without consent of the black male participants who were offered no treatment once the disease occurred although syphilis had become curable in the later stages of the study. President Clinton issued a formal Presidential Apology, and after an investigation, a National Center for Bioethics in Research and Health Care at Tuskegee University was established. This is just one example of many racial and eugenically motivated forms of medical exploitation of Indigenous and disabled peoples around the world violating their rights and dignity.

References

Alac, M. 2008. Working with Brain Scans: Digital Images and Gestural Interaction in FMRI Laboratory. *Social Studies of Science*, 38(4), pp. 483–508.

Baccarini, Elvio and Malatesti, Luca. 2017. The Moral Bioenhancement of Psychopaths. *Journal of Medical Ethics*, 43(10), pp. 697–701.

Bigenwald A, Chambon V. 2019. Criminal Responsibility and Neuroscience: No Revolution Yet. *Frontiers Psychology* 10, p. 1406.

Butler, T., Schofield, P. W., Knight, L., Ton, B., Greenberg, D., Scott, R. J., Grant, L., Keech, A. C., Gebski, V., Jones, J., Ellis, A., Weatherburn, D., Wilhelm, K., Jones, A., Churchill, A., Allnutt, S., Mitchell, P. B., Chappell, D., Deste, C., Villa, D., and Carr, V. 2021. Sertraline Hydrochloride for Reducing Impulsive Behaviour in Male, Repeat-Violent Offenders (ReINVEST): Protocol for a Phase IV, Double-blind, Placebo-Controlled, Randomised Clinical Trial. *BMJ Open*, 11, pp. e044656. https://doi.org/10.1136/bmjopen-2020-044656.

Chang, Z., Lichtenstein, P., Langstrom, N., et al. 2016. Association Between Prescription of Major Psychotropic Medications and Violent Reoffending After Prison Release. *JAMA*, 316(17), pp. 1798–1807.

Dumit, Joseph. 2004. *Picturing Personhood: Brain Scans and Biomedical Identity*. Princeton: Princeton University Press.

Dumit, Joseph. 2012. Critically Producing Brain Images of Mind. In Choudhury, S. and Slaby, J. (eds.), *Critical Neuroscience: A Handbook of the Social and Cultural Contexts of Neuroscience*, pp. 195–225. Malden: Wiley-Blackwell.

Farahany N. A. 2016. Neuroscience and Behavioral Genetics in US Criminal Law: An Empirical Analysis. *Journal of Law and the Biosciences*, 2(3), pp. 485–509

Farahany, N. A. 2015. Neuroscience and Behavioral Genetics in US Criminal Law: An Empirical Analysis. *Journal of Law and Biosciences*, 2, pp. 485–509.

Federle, K. H. and Skendelas, P. 2009. Thinking like a child: Legal implications of recent developments in brain research for juvenile offenders. In Freeman, M. D. and Goodenough, O. R. (eds.), *Law, Mind and Brain*, pp. 199–214. London & New York: Routledge.

Garland B, ed. 2004. *Neuroscience and the Law: Brain, Mind and the Scales of Justice*. New York: Dana Press.

Garland, B. and Frankel, M. S. 2006. Considering Convergence: A Policy Dialogue About Behavioral Genetics, Neuroscience, and Law. *Law and Contemporary Problems*, 69(1/2), pp. 101–113.

Greely, H. T. and Farahany, N. A. 2019. Neuroscience and the Criminal Justice System. *Annual Review of Criminology*, 2, pp. 451–471.

Heinemann, L. and Heinemann, T. 2010. Optimise your brain!–Popular science and its social implications. *BioSocieties*, 5, pp. 291–294.

Hodgson, D. 2000. Guilty mind or guilty brain? Criminal responsibility in the age of neuroscience. *Australian Law Journal*, 74(10), pp. 661–680.

Jasanoff, Sheila. 2015. Serviceable Truths: Science for Action in Law and Policy. *Texas Law Review*, 93(7), pp. 1723–1749.

Kahn, J. 2015. Neuroscience, Sincerity, and the Law. *Bergen Journal of Criminal Law & Criminal Justice*, 3(2), pp. 203–220.

Kudlek, Karolina. 2022. Towards a Systematic Evaluation of Moral Bioenhancement, *Theoretical Medical Bioethics*, 43(2–3), pp. 95–110.

Lichtenstein, P., Halldner, L., and Zetterqvist, J. 2012. Medication for Attention Deficit-Hyperactivity Disorder and Criminality. *New England Journal of Medicine*, 367, pp. 2006–2014.

Mackenzie, Adrian, and Roberts, Celia. 2017. Adopting Neuroscience: Parenting and Affective Indeterminacy. *Body & Society*, 23(3), pp. 130–155.

Margulies, D. S. 2012. The Salmon of Doubt: Six Months of Methodological Controversy Within Social Neuroscience. In Choudhury, S. and Slaby, J. (eds.), *Critical Neuroscience: A Handbook of the Social and Cultural Contexts of Neuroscience*, pp. 273–285. Malden: Wiley-Blackwell.

Mascolo, J. and Bui, J. 2023. Chemical and Surgical Castration for Sex Offenders. *FindLaw*, https://www.findlaw.com/criminal/criminal-charges/chemical-and-surgical-castration.html

Morse, S. J. 2006. Moral and legal responsibility and the new neuroscience. Neuroethics: Defining the issues in theory, *Practice, and policy*, pp. 33–50.

Morse, S. J. 2011a. Neuroscience and the Future of Personhood and Responsibility. In J. Rosen and B. Wittes (eds.), *Constitution 3.0: Freedom and Technological Change*, pp. 113–129. Washington, DC: Brookings Institution Press.

Morse, Stephen J. 2011b. Lost in Translation? An Essay on Law and Neuroscience. *Law and Neuroscience,* 13, pp. 529–562.

Morse, Stephen J. 2014. Neuroethics: Neurolaw. *The Oxford Handbook of Topics in Philosophy* (online ed., Oxford Academic, April 1). https://doi.org/10.1093/oxfordhb/9780199935314.013.45. Accessed 6 August 2023.

Morse, Stephen J. 2016. Actions speak louder than images: The use of neuroscientific evidence in criminal cases. *Journal of Law and the Biosciences*, 3(2), pp. 336–342.

Morse, Stephen J. 2017. Neuroethics: Neurolaw. *The Oxford Handbook of Topics in Philosophy*. https://doi.org/10.1093/oxfordhb/9780199935314.013.45. Accessed 6 Mar. 2024.

O'Conner, C., Rees, G., and Joffe, H. 2012. Neuroscience in the public sphere. *Neuron*, 74, 220–226.

Pernu, T. K. and Elzein, N. 2020. From Neuroscience to Law: Bridging the Gap. *Frontiers in Psychology*, 11, p. 1862.

Pickersgill, M. 2013. The Social Life of the Brain: Neuroscience in Society. *Current Sociology*, 61(3), 322–340.

Pickersgill, M. 2011. Connecting Neuroscience and Law: Anticipatory Discourse and the Role of Sociotechnical Imaginaries. *New Genetics and Society*, 30(1), pp. 27–40.

Pitts-Taylor, V. 2016. *The Brain's Body: Neuroscience and Corporeal Politics*. Durham, NC: Duke University Press.

Rees, T. 2016. *Plastic reason: An anthropology of brain science in embryogenetic terms*. University of California Press.

Reilly, P. R. 2015. Eugenics and Involuntary Sterilization: 1907–2015. *Annual Review of Genomics in Human Genetics*, Vol. 16, 351–368.

Rose, N. 2010. 'Screen and Intervene': Governing Risky Brains. *History of the Human Sciences*, 23(1), pp. 79–105.

Rose, N. and Abi-Rached, J. M. 2013. *Neuro: The New Brain Sciences and the Management of the Mind*. Princeton, NJ: Princeton University Press.

Sánchez-Torres, G., Leal, N. and Pino, M. 2024. *Data-Driven Analysis of MRI Scans: Exploring Brain Structure Variations in Colombian Adolescent Offenders*. Data 2024, 9, 7.

Sapolsky, Robert. 2004. The Frontal Cortex and the Criminal Justice System. *The Royal Society*, 359(145), p. 1787.

Settumba, S., Butler, T., Schofield, P., Chambers, G. M., and Shanahan, M. 2021. Societal Preferences for the Treatment of Impulsive-Violent Offenders: A Discrete Choice Experiment. *BMJ Open*, 11. https://doi.org/10.1136/bmjopen-2019-033935.

Shen, F. X. 2016. The Overlooked History of Neurolaw. *Fordham Law Review*, 85, pp. 667–695.

Sirgiovanni, Elisabetta and Garasic, Mirko Daniel. 2020. Commentary: The Moral Bioenhancement of Psychopaths. *Frontiers in Psychology*, 10, p. 2880.

Slaby, J. and Choudhury, S. 2018. Proposal for a Critical Neuroscience. In Meloni, M., Cromby, J., Fitzgerald, D., and Lloyd, S. (eds.), *The Palgrave Handbook of Biology and Society*, pp. 341–370. London: Palgrave Macmillan.
Spenser, K., Bull, R., Betts, L., et al. 2019. Executive Functioning as a Predictive Measure of Offending Behaviour. *Journal of Criminal Psychology*, 9(1), pp. 10–22.
Taylor, J. S. 1995. Neurolaw: Towards a New Medical Jurisprudence. *Brain Injury*, 9(7), pp. 745–751.
Taylor, J. S. 1997. *Neurolaw: Brain and Spinal Cord Injuries*. Houston: Thompson Legal Publishing.
Vidal, Fernando. 2009. Brainhood, Anthropological Figure of Modernity. *History of the Human Sciences*, 22(1), p. 7.
Vincent, Nicole A. 2013. *Neuroscience and Legal Responsibility*. Oxford: Oxford University Press.
Wilson, E. A. 2015. *Gut feminism*. Duke University Press.

6

Babies: Kinship and Relations

For all the relational inventiveness that Euro-Americans pour into their systems of knowledge, or the work that goes into making society, or the passion of a judge's plea that one-time parents give heed to their relation with a child, the law does not recognise a relationship as a legal subject. Only individuals (individual persons) can be legal subjects.

Marilyn Strathern, *Kinship, Law and the Unexpected*

The routines of family life have usually meant that relationships without responsibilities tend to fade away. A truism about knowledge can keep them in view: the genes that carry the data informing you what you are at the very same time comprise the mechanisms that have the potential to bring about what you are. This looks like a reworking of an old theme, the constitutive nature of kinship knowledge. But to find kinship knowledge in the gene is, so to speak, to find it in itself. Knowledge and kinship become momentarily inseparable.

Marilyn Strathern, *Kinship, Law and the Unexpected*

S. van Wichelen and M. de Leeuw, *Biolegality*, Biolegalities,
https://doi.org/10.1007/978-981-99-8749-8_5

Introduction

In February 2015, the UK passed legislation allowing the creation of "three person babies" via a new technology called Mitochondrial Replacement Therapy (MRT) (Gallagher 2015).[1] The technology inserts the nuclear material from the intended mother with defective mitochondrial DNA into a donated oocyte with healthy mitochondria. This prevents the transfer of serious mitochondrial diseases—such as Kearns-Sayre syndrome or Leigh's syndrome—to the child.[2] While the technique involves genetic material from the donated oocyte, the egg retains the nuclear DNA from the intended mother. The technique thus opens the possibility of three genetic parents—mother, oocyte donor, and father—prompting a public and legal debate in a range of jurisdictions, including Mexico where the procedure was first successfully performed in 2016.

The possibility of two *genetic* mothers (or three *biological* mothers when the child is delivered via gestational surrogacy) destabilizes the legal foundations of maternity and paternity. Since its inception in Roman law, legal motherhood is biologically defined and founded on the principle of birth evinced by the Latin adage *mater cemper certa est* (the mother is always certain). Legal fatherhood, on the other hand, was long based on the social contract of marriage, though this has changed too, with the widespread use of paternity testing. MRTs, and reproductive technologies more generally, scramble these legal conventions.[3] As genetic parentage is becoming more and more the norm to adjudicate children born from assisted reproduction (including surrogacy), paternity is becoming more certain than maternity. In this chapter, we explore these shifts in human reproduction and examine their consequences for society and culture.

The debate around MRTs echoes longstanding concerns about the legality of reproductive technologies and their impact on notions of parenthood, kinship, and heredity. Assisted reproductive technologies (ARTs) include a wide range of applications. In its widest understanding ARTs include conceptive technologies such as Intracytoplasmic Sperm Injection (ICSI) and In Vitro Fertilization (IVF), IVF use in gestational surrogacy, and the technologies and therapies that observe, diagnose, and repair gametes and embryos, including ultrasounds, MRTs, and

Preimplantation Genetic Diagnosis (PGD). ARTs do not function in a social vacuum and are informed by reproductive markets constituted by the political economies of oocyte, sperm, and embryo donation, the availability of surrogates, and the regulatory frameworks of these practices. While the bioethics literature on reproductive technologies largely focuses on the personal ethico-moral dimensions of ARTs, our attention here is on how the biolegal arrangements resulting from ARTs redefine biological and social relations. Three things are central to such legalities: (1) the ways genetic knowledge shapes the legal understanding of parenthood, (2) the practices, techniques, and habits of law that rearticulate these new knowledges, and (3) the global reproductive markets that steer sovereign states in new directions.

The Uncertainty of Motherhood

Ever since Robert Edwards and Patrick Steptoe invented In Vitro Fertilization (IVF), the technique allowing conception outside of the womb, infertility ceased to be an endpoint, but instead the start to a range of treatments and technologies addressing infertility. The 1978 births of Louise Edwards in the UK, and baby Durga in India, showed not only the possibility that modern biosciences could create life by intervening in natural procreation, but also proved social and ethical acceptability of so-called "test tube babies." Justified as treatment for infertility—thereby framing infertility as disease—IVF was quickly normalized and integrated into medical practice, though with varying degrees of financial assistance from the state.[4] As a technique, IVF was just the start of reproductive techniques developed in the wake of Louise and Durga's births and the spark igniting a series of heated confrontations with law. The most important of these controversies is the introduction of third parties in reproductive practices, which complicates the legality of reproduction. Donated sperm, donated oocytes, and contract-based surrogacies destabilizes parenthood in various, contractually mediated, ways.

Perhaps the most compelling catalyst for legal issues involving reproductive technology is the case of gestational surrogacy. In the US, where certain states have liberal surrogacy regulations, high profiled supreme court cases set the tone in qualifying rightful parenthood in such cases. For instance, in the famous *Johnson versus Calvert* case, IVF allowed the gestation of non-genetically related children, leading to the dismissal of the surrogate's parental claims. Instead, the court ruled that legal motherhood should be assigned to the (contractually) "intended" mother. This judgment overruled the earlier high-profile case of Baby M,[5] where a surrogate received parental visitation rights based on the principle of *mater cemper certa est*. Prior to IVF and the use of donated oocytes, this legal convention based on the process of gestation and the event of birth prevailed in any case adjudicating motherhood; the advent of IVF and the donation of oocytes, put this principle into question, undermining its claim as a presumption of law and by law (*praesumptio iuris et de iure*). While not always explicit, this revision of the law frequently leads to a new status quo, particularly in cases of gestational surrogacy where the father now becomes the more certain parent (Melhuus 2009; D'alton-Harrison 2014). The UK for instance, recognizes paternity through DNA testing, while this is not the case for maternity, an uneven, and inverted, recognition of parenthood. In contrast to the intended mother (with or without a genetic connection), the intended father in a surrogacy arrangement has a genetically privileged position through DNA testing.[6] Scholars (Almeling 2011; Purvis 2012; Schiff 1994; Strathern 1996; Thompson 2009) argue that the turn to legal paternity results in valuing donated sperm over egg donation; the first can automate legal parenthood, while the latter cannot.

Curiously, it is not only the rise of genetic parenthood (and therefore biological parenthood) over forms of social parenthood that dominates the legal debate around parentage in the age of reproductive technologies. While genetics call forth new forms of legal recognition and social belonging, the notion of "intended" parenthood is gaining legal and social traction, as is illustrated well by the Californian *Re Marriage of Buzzanca* case. There, a married couple, using a donated embryo and a surrogate, produced a reproductive arrangement,[7] involving five parties: the two genetic parents donating the embryo, the gestational parent (the surrogate), and the intentional parents (with no genetic connection to the child whatsoever). Before the child was born the intending parental couple split and the intentional father denied any parental responsibility for the child. The court ruled that it was the couple's (medical) decisions that brought the child into being; without those intentional actions, the child would not exist. By assigning parental responsibility to the intentional parents, the court's ruling prioritized contractual evidence over genetic evidence, which begs the question of how the law relates to the social vis-à-vis the biological. As law is itself always already social, the court's decision articulated this position; "the law's concern for the social order overrides what science deems to be the facts of the matter" (see p. 334 in Jasanoff 2006).

The court's seemingly paradoxical insistence on the genetic *and* social relevance of parenthood suggests the active work of legal reasoning in matters of biology and life. In many paternity cases, the genetic link is primary in establishing paternity, giving genetic fathers automatic legal parentage, while intentional motherhood takes precedent over gestational motherhood. As the STS scholar Sheila Jasanoff explains (2001), the reasoning that occurs in and outside of courts points to ideas of what is deemed natural and normal. Before IVF, natural motherhood accorded with legal motherhood based on gestation, which grounded the principle that the mother is always certain. With the advent of gestational surrogacy, what is "natural"—and therefore what is legal—is now up for debate, troubling the underlying conception of normality and naturalness (see p. 45 in ibid.).

Our opening example of mitochondrial replacement techniques exemplifies the further complicating impacts of bioscience on parentage, where a phenomenon like multiple genetic mothers is compounded by the discovery of other new facts about life during pregnancy that further blur the distinction between genetics and biology, for instance between the hereditary and the gestational. Emerging knowledge from epigenetics, the microbiome, and microchimerism in immunology, suggests that the mother–fetus relationship (genetically and gestationally) is much more complex and entwined than previously thought, complicating ideas around genetic stability (Hird 2007; Richardson 2015; Shildrick 2022; Van Wichelen 2016), but also of race and belonging (Vora 2015; Van Wichelen and Keaney 2022; Keaney 2022; Van Wichelen 2022). Such postgenomic research therefore directly or indirectly affects how we view, value, and legally recognize motherhood in relation to the biological contribution of the gestating woman. The feminist technoscience scholar Margrit Shildrick (2022), for instance, describes how microchimerism may have serious implications in the ethico-legal context. Focusing on the now widely recognized existence of maternal-fetal microchimerism, which establishes "the diversity of DNA coding in a singular body," the implications for law can be substantial given how society has aligned genetic relatedness with the identification of legal parenthood and social relations of kinship. Challenging conventional ideas around the stability of genetic identity, the microchimeric destabilization of genetic origins raises the question whether our ethical and legal apparatus are sufficient to respond to capture the complexity of this emerging science and their reconfiguration of the legal person.[8]

Banking Time, Freezing Life

While the literature on gestational surrogacy points us toward the contemporary idiosyncrasies of legal motherhood, recent social science scholarship on reproductive cryobiology, including the freezing of gametes (sperm and eggs) and embryos, direct us to notions of biological time and the legal problems that arise when biological time is frozen, as it were, by cryobiological technologies.

Cryobiology is not new, having been introduced in the 1950s for use in animal husbandry to boost and coordinate reproduction (Jackson 2018; Lockwood 2011; Waldby 2015). It was not until the past decade, however, that we witnessed its impact on human reproduction, and their consequences in society and law. Novel techniques of vitrification allow much longer preservation of gametes and embryos (or zygotes) without loss of vitality. In the past decade, clinical trials show that the use of frozen sperm and eggs does not lead to a significant reduction in the chance of healthy offspring, allowing governmental and medical bodies to move the practice from experimental and medical use to commercial and non-medical spaces (see p. 472 in Waldby 2015). Because of the limited viability of oocytes after the age of 35 (colloquially referred to as the "fertility cliff"), single women are targeted as a growing market for egg freezing by biobank companies and fertility clinics. The success of these new vitrification technologies, then, pause and stretch biological time more successfully than in the past, leading to a vast growth in women banking their eggs for social rather than medical reasons.

Sociologist Catherine Waldby's (2015) empirical research on oocyte freezing describes these women's choices to bank their eggs as guided by the idea of genetic continuity. Egg freezing situates women in what she calls "generational time," where choices to harvest and freeze eggs are motivated by intergenerational appeals to genetic continuity (see pp. 478–479 in Waldby 2015). Whereas publics negotiate the necessity of "delaying" motherhood, markets promote egg banking as a purchasable service for women hoping to "extend" the possibility of motherhood (Van de Wiel 2015). Egg banking functions like an insurance policy, acting as an extended guarantee of motherhood (Lockwood 2011; Waldby 2015). This analogy illustrates the ethical and legal issues that arise out of these practices.

While publics generally agree that the use of gamete freezing for medical purposes (such as chemotherapies in cancer treatments) is legitimate, the so-called "social freezing" of eggs—for what is seen as the postponement of motherhood—is up for ethical debate (Bernstein and Wiesemann 2014; Lafuente-Funes 2023; Van de Wiel 2020). In many jurisdictions the expansion of reproductive industries and private biobanks catering to egg freezing have generated calls for regulation, framing the social problem of managed infertility within a consumerist and neoliberal understanding of reproductive freedom and autonomy. Legal questions emerge when considering the boundaries of this freedom, many of which are framed in property questions (i.e. who do the oocytes belong to and what are the terms of ownership) but also in questions concerning the "best interest of the child" (should there be age limits to women's conception).

New practices of sperm retrieval and frozen oocytes have led to a tremendous rise in cases involving posthumous reproduction. Use of gametes, which can be frozen prior to death or retrieved postmortem, raises paternity issues in cases where a deceased partner's or spouse's sperm is used for artificial insemination resulting in live birth. The legal questions that emerge pertain to whether legal parentage is assigned to the late spouse based on genetics, or whether intention is needed to confirm paternity, as well as to whom paternity should be assigned to when the mother remarries (Zafran 2007)? In a high-profile British case starting in the late 1990s, Diane Blood fought the Human Fertilisation and Embryology Authority for the control over her late husband's sperm so that she could travel to another country where artificially inseminated posthumous conception is allowed.[9] She won the case, and in 2003, won another case to record her late husband's name on the birth certificates of her two sons born from IVF techniques.

Court cases in the arena of oocyte freezing followed more than a decade later. In 2015, a 59-year-old woman called "Mrs. M" launched a legal battle in the UK for the control of her late daughter's frozen eggs (who passed away from colon cancer when she was 28 years old) in order to proceed with fertility treatments in the US to carry her own grandchild.[10] The *Mr and Mrs M versus Human Fertilisation and Embryology Authority* case challenged the independent Human Fertilisation and Embryology Authority's refusal—based on the absence of an informed consent form—to release Mrs. M's daughter's eggs. Mr. and Mrs. M appealed to the court, claiming they wanted to grant their daughter's wish to procreate by bringing her eggs to an American fertility clinic for the mother, Mrs. M, to undergo the IVF procedure with donor sperm to act as a gestational surrogate for their late daughter's child. In 2016 the UK supreme court ruled in favor of the couple, deciding that HFEA's refusal failed to fully consider the daughter's wishes for the posthumous use of her eggs (Bowcott 2016). This judgment opens the way for a broader discussion of egg freezing's variegated use beyond the common notion of personal use. While the NHS in the UK and Medicare in Australia have guidelines and restrictions, including age limits, for determining eligibility for subsidized treatments, many countries do not have legal age limits or other restrictions on fertility treatments.

The lack of laws for gamete freezing produces considerable uncertainty for the people involved and prompting calls for regulation (Cahn 2009). However, even where governmental and clinical guidelines and procedures (often related to age) exist (such as those at the NHS in the UK), lack of clear international laws and standards leave practices of freezing gametes and embryos to the vagaries of the market. Recent trends reveal that US women increasingly travel to freeze their eggs.[11] This is because in the US a procedure to retrieve and store eggs can be around fifty to two hundred percent of their average annual income.[12] New companies attract customers who are combining egg freezing procedures with a vacation in Mexico, Barbados, or Spain in their bid to offer more affordable packages compared to what they can get back home. The global fertility tourism market—estimated at USD 115.6 billion in 2022—is growing significantly and expected to rise by thirty percent over the next seven years.[13] Scholar of social medicine Lucy van de Wiel argues therefore that other than changing the temporality of biological fertility, cryopreservation of oocytes also financializes the procurement and distribution of oocytes, making them a lucrative business. Regulation can curb this development, but can also facilitate it depending on how these bilateral, multilateral, or international developments of cross-border rules and standards develop.

"One of Your Own": Genes Are Us

In all the cases above, the specter of genetics runs deep in the procreative desires and hopes reproductive technologies fuel. The popularity and immense growth of IVF signals the cultural significance of genetics in family-making. Since its arrival, critical commentators have pointed to the negative consequences of IVF's development on practices of social family-making such as adoption and fostering. More recently, scholars studying kinship in the Anthropocene are calling for an expanded idea of kinship that includes non-human and multi-species forms of family-making (Haraway 2016; Kirksey 2014).

Law is also witnessing a displacement of the Euro-American nuclear family—"natural" father, mother, and children—as model for legally recognized families. This is because reproductive technologies challenge modern family law's dependence on the cultural, biological, and legal template of heterosexual, natural, and conjugal procreation. In that regard, Marilyn Strathern (2005) conceptualizes what she calls "dispersed kinship" to describe and analyze the multiple parties involved in making families in the context of assisted reproduction. As the previous sections indicated, apart from the intended parent/s, numerous technologies, institutions, and other people play key roles in "dispersed kinship"; as gamete donors, clinics, doctors, ultrasounds, hormone therapies, extended family members, medical and state authorities, etc. These entities may, depending on the assisted reproductive arrangements, become party to legal claims, especially when things go awry as was often the case with gamete freezing and surrogacy throughout the 2010s.

It is remarkable that, on the one hand, genetic connection forms a key determinant in the use of reproductive technologies, pointing to the cultural centrality of genetic continuity and belonging, while, on the other, many IVF cases use (anonymous) donors whose genetic contribution is downplayed or ignored. This paradoxical situation presents itself in court too. Legal scholars note the way judges and court rulings tend to stress the genetic link for the intended parent/s in granting them legal parentage, while at the same time disregarding the genetic contributions of donors. These contradictions have much to do with how we are rearranging kinship knowledge at a time when both genetic connection and social relatedness have gained more traction.

Accessibility of genetic information and manipulation not only complicates legal parenthood, but also kinship knowledge more broadly. The surge of genetic testing in recent decades has raised societal and bioethical concerns about the affect of such testing regimes on individuals, families, and the communities. Genetic information, with its potential to reveal new kin or sever assumed kinship, challenges the basic premise of individual autonomy in bioethics (Gibbon 2002; Lebner 2000). Genetic information itself becomes "constitutive" as it is coupled with the social and cultural idea of identity: "kinship knowledge is a particular kind of knowledge; the information (and verification) on which it draws is constitutive in its consequences" (see p. 68 in Strathern 1999, see also Carsten 2000). While individuals actively manage and negotiate this information, rather than blindly acting on it, once it is known it cannot be "unknown" and its affect becomes permanent and irrevocable.

The move in many countries against anonymous donorship in domestic policies for assisted reproduction and surrogacy attests to the view that the genetic relationship between the oocyte donor and the child is more important to identity than previously thought. Translated into the language of rights and international law, the geneticization of identity establishes the doctrine of "the right to know." Although the debate surrounding anonymity in donor-conceived conceptions remains highly contentious within bioethics, from a human rights perspective the "right to know" is firmly entrenched in international conventions as well as domestic laws. This is especially notable in relation to adoption, where the idea of replacing secret adoptions with "open adoption" is gaining traction in Northern America, Europe, and Australia (Modell 2002; Howell 2009; Yngvesson 2010). A similar trend discouraging or even barring anonymity in sperm and egg donations is catching on in more and more jurisdictions, strengthening the directive toward openness within national jurisdictions. These developments reveal how understandings of relatedness change over time.

The rise and prominence of genetics and biotechnology is actively reworking the meaning of "blood relationships." Propelled by the growing Euro-American interest in genetic origins (Nash 2002, 2004), adoptees and donor-conceived children increasingly depend conceptually and commercially on the global "kinning" of genetic connectivity (Howell 2003). Here, people who find genetic connection—through for-profit DNA testing companies—progressively turn those connections not only to kin relations, but also to new (social and cultural) community frameworks in which to understand their (true) Self. The geneticization of human identity collapses identity, and by proxy kinship, into biology, advancing, not only a discourse of determinism, but also the constitution of political and legal subjectivities that draw on genetic testing to support their claims in court.

Rights and Interests

Within legal contexts, lawyers and judges utilize the language of rights to deal with the constitutive effects of biological or biomedical information on family life. In other words, courts, judges, and lawyers extended certain understandings of the role of human genetics over others when integrating biological knowledge and controlling what this knowledge reveals about humanity (see p. 2 in Van Beers et al. 2014). Within the context of reproductive technologies, for instance, this has led to an extension of children's rights in the past few decades, especially the "right to know." The idea of "the right to know" encompasses several different things in reproductive practice: the right to know one's biological (genetic) origins; the right to access medical information from genetic parents/donors (including information about genetic predispositions and diseases); and, in the case of adoption, the right to know one's ethnic, religious, and cultural background. While the right to know has gained value in Euro-American legal contexts since the 1970s, it was only formally recognized two decades later, by the Convention for the Rights of the Child (CRC), which asserts a vital interest in children knowing their biological ancestry (see p. 138 in Besson 2007).[14]

Nonetheless, several legal scholars (Besson 2007; Diver 2014), have identified inherent paradoxes in the "right to know" provisions of international law. For instance, Article 16 of the CRC protects the privacy of the family, which in the case of some member states gives legal preference to the privacy of biological parents/donors, thus contravening the child's "right to know" the identity of their biological parents (see p. 87 in Diver 2014). Moreover, as Diver describes in relation to Article 21 of the CRC, which explicitly covers the practice of intercountry adoption, "the provision is effectively silent on the issue of whether biogenetic information ought to be made available at a later stage, largely leaving the matter to be decided by domestic legislators and jurists" (see p. 80 in Diver 2014). While these separate clauses to the CRC complicate an unequivocal "right to know" within domestic jurisdictions, the normative weight of the "right to know"—which does not seem to carry the inconsistencies—remains strong in public discourse.

Besides children's rights, commentators observe that reproductive technologies expand reproductive rights to include procreative rights, which can be simply understood as the right to have children as opposed to the right to choose when and how to have children (Mutcherson 2016; Van Wichelen 2019a). Although procreative rights do not exist in the formal letter law, individuals and communities—from both conservative and progressive corners—cite the framework of rights in struggles over reproductive choice, freedom, and liberty. Depending on the legal framework of a particular jurisdiction, reproductive technologies can be directed into a more conservative agenda, where it serves to consolidate the traditional and nuclear family. Take, for instance, the common practice of automatically bestowing legal fatherhood on husbands, which, in an IVF situation involving donor sperm, opportunely obliterates genetic claims to parental rights. On the other hand, reproductive technologies are a vehicle by which LGBTIQ individuals and communities assert their equality in the domain of reproduction, often calling upon procreative rights to enact the same liberties as heterosexual individuals. For instance, legal scholars continue to debate whether such rights of procreation should be approached as negative or positive rights (Roberts 1995; Robertson 1995). And this question then implicates the role of the state and to what extent the state is obligated to assist citizens with their reproductive choices, including by means of legislation (see p. 954 in Rao 1995).

In addition, the framework of rights often conflicts with the framework of interests, as is the case with the "best interest of the child" so frequently invoked in domestic and international courts. As many commentators argue, though, this principle is ambiguous and too commonly depends on the discretion of decision makers—whether in courts, clinics, or child welfare institutions. This discretion frequently occurs in situations where the state intervenes in family life, such as in the case of cross-border surrogacy arrangements. Here intended parents, upon returning with a child born out of a commissioned surrogacy arrangement abroad, are confronted with a state that takes control of the child based on its prohibition against commercial or cross-border surrogacy.

The cases *Mennesson v. France* and *Labassee v. France* are landmark cases in this regard.[15] In this double case of cross-border gestational surrogacy from the US to France—with a genetic link between the intended father and child—the 2014 ruling held that France, by not acknowledging legal parenthood and not providing the children French citizenship, intervened in the private life of the children, if not necessarily the parents. Moreover, because family life was well-established between the intended parents and the children (the children were thirteen and fourteen at the time of the hearings and lived with their intended parents since birth) the court determined it was in the best interest of the children to recognize parenthood and to grant the children legal identities and citizenship.

Globalization and Conflicts of Law

What does it mean when engineered biology (in the form of genetic material) can travel across borders, whereas social, "intentional," or de facto relations cannot? How do we regulate "private family life" in the broader context of the social and the global?

Scholars of cross-border reproductive services target the disruptive nature of reproductive industries, including its entanglement with post-Fordist capitalism, digital technologies, and lack of global regulatory oversight (Whittaker 2018; Mitra and Schicktanz 2018). Biolegalities are central to the international choreography of these industries, both facilitating and mitigating the disruptions they cause. For instance, because many jurisdictions already have surrogacy laws, with clear prohibitions formed through consensus and expert opinion, the popular "law-lag idea"—the ways law follows technology rather than preceding it—does not account for the relation between legal regulation and private international reproductive practices. Nonetheless, children from surrogacy arrangements still move from countries with liberal regulations to those with stricter ones. That is, global reproductive practices, like surrogacy, do not happen in a vacuum; instead they take place in the context of other legalities, orders, and standards, that, even if they do not necessarily relate to the phenomenon in question, direct activity toward legal acceptance.

Take, for example, the surrogacy regulations in Australia, where there are different regulations in each domestic state. New South Wales (along with three other states) prohibits commercial surrogacy and criminalizes cross-border surrogacy with clear laws banning the practice of transnational surrogacy. Yet, commercial gestational surrogacies from places like India and Thailand have occurred in the past and the children legally reside in Australia, including in those states where the practice is explicitly banned (Whittaker 2018; Van Wichelen 2019b). The complex (il)legality determining how these circulations take place is a form of biolegality. What happens is that the genetic link between an Australian intended parent and the child automatically results in Australian citizenship for the child. Despite the child being gestated outside of the national jurisdiction, paternal genetics grants these children citizenship. So even though commercial surrogacy is banned, citizenship laws allow parentage relations to be established on the basis of genetic connection. Moreover, by the time some of these cases are challenged and go to court, the child or children will have lived a considerable time with the intended parents. Often this means that courts find it is in the best interest of the child to stay with their intended parents and give them shared parental responsibility.

The lack of a clear and coherent global regulatory framework is often blamed for the disruptive effects of international reproductive industries, suggesting the need for transnational regulation of reproductive technologies (see Dearle 2015), particularly in the arena of cross-border surrogacy arrangements (Trimmings and Beaumont 2011). Transnational regulation in the field of private, rather than public, international law is seen, then, as a panacea for resolving the conflicts that grow out of reproductive markets.[16] Legitimating forces and justificatory frameworks, both private and public, are part and parcel to the operation of law in transnational regulation, but what else does regulation "do"? The socio-legal scholar Emilie Cloatre argues that there are "silent ways in which law may work through the travels of materials that are already loaded with meanings, yet go on to, relationally, 'do their own thing' with it" (see p. 657 in Cloatre 2018). Beyond the framework of legitimacy lies the production of knowledge and the knowledge effects of regulation. Put differently, besides the production of legitimating orders, regulation and law also produce "a set of institutions, actors, doctrines, ideas, [and] material documents [of] knowledge practice" (see p. 207 in Riles 2008). These practices often do not cohere in ways that connect parts of the global network where different truths are being produced and practiced. Instead, as the legal and STS scholar Annelise Riles argues, they represent an "anti-network" in which practices under the umbrella of private international law form unstable and contingent arrangements. This instability on the level of practice, however, stands in stark contrast to the stabilization of legitimacy. Procedures, routines, and bureaucracies allow for the dissolution of normative controversiality through a depoliticizing process; the contradiction between the instability of the network becomes resolved through legitimacy. A reason why regulations unfolding across a contingent plane are granted legitimacy comes down to the working of power and inequality in the global legal system represented here through the relation between private and public law. In the world of fertility markets, private law—originating in the powerful Global North—overrides public law in the Global South, and yet does so in ways that produce legitimacy. The legal scholar of private international law—Horatia Muir Watt—argues that private international law has always been about enabling rather than disciplining capacities (2011: 354) and calls for a global legal pluralism to reconsider the conflicts of law system (see also Watt 2016). Following Gunther Teubner (2012), she proposes a societal constitutionalism "for unbounding the conflict of laws, which might then emerge as a form of de-centred, reflexive coordination of global legal interactions" (see p. 313 in Watt 2016). Although Watt does not address the specificity of reproductive markets, the globalization of the reproductive economy is exemplary in stipulating the workings of contractual freedom, a key feature in the global political economy of private ordering. Approaching this problem through societal constitutionalism will attend to justice (rather than contract) and balance markets against other values.

Conclusion

The emergence and globalization of reproductive technologies, and their attendant effects on how we understand kinship, produce novel problems for legal thinking. Rather than assuming that law lags behind, or that future regulation will restore a previous equilibrium, it is more accurate to identify an active reorganization of legal and biological knowledge occurring, which displaces—modifies, puts on hold, or otherwise affects—conventional understandings of kinship. By activating and mediating different legal domains with their own knowledge systems, law contributes to this remodeling of kinship. In society, such remodeling comes out of our legal engagements with who defines and how we define our kin. Biology or nature, and our understanding of them, play a role but so do technologies, including legal technologies. As science studies scholar Sarah Franklin notes, technology and biology produce new classifications of "biological relatives" (2013). Translated to law, these biolegal exchanges similarly produce new forms of "legal relatives."

Observing the trends in reproductive technologies, we witness the reordering of life as much as the reshaping of society. While kinship is remade and relearned through these biological technologies, cultural forms of reproduction—such as egg freezing—can be thought of as "cultured-up technology" (see p. 4 in Franklin 2013). Here, conservatism or traditionalism is not (necessarily) opposed to technology; rather, as Rabinow explains, conservatism can adopt technoscience itself to maintain or reify traditional notions of kinship (1998). This is why partisans of pro-life and pro-choice positions both draw on narratives of assisted reproduction, as much as pro-natalist and feminist ideologies drive reproductive industries and legalities.

According to Hannah Landecker (2005), reproductive biotechnology not only reorders human life, but it also reorders biology itself. The ability to preserve gametes beyond the lives of the donors, for instance, suspends biological time, forcing society to reconsider (consciously or not) the relationship between generational time and hereditary kinship. This does not occur without a socio-legal discourse about the ethics of these practices (Kroløkke and Adrian 2013). Our main point here is that the reordering of life is not merely technological or scientific. Law plays a key role, as does commerce. Returning to the case we opened with, a technology like mitochondrial replacement techniques may create children with three genetic parents, the legal standing of a three-parent baby, however, depends on the social shaping of this biological fact. Similarly, while technology now enables postmortem biological reproduction, the assignment of legal parentage to deceased men and women is a matter of law, not science. Where there is no such legitimacy, postmortem parenthood—as a set of enforceable rights and responsibilities—does not exist. Ultimately, then, biotechnology not only affects individuals but impacts communities at large, making it fundamentally a question of social responsibility. In the next chapter, we probe these encounters between biology, law, and communities, and tackle some of the biolegal puzzles that emerge from them.

Notes

1. Mitochondrial Replacement Technique (MRT) is the most common definition but according to some the correct term is human nucleus transference (Van Beers 2018). We will stick to MRT, since it is the generally used term in the scholarship.
2. Kearns-Sayre and Leigh syndrome are rare disorders. The first presents itself by droopy eyelids while the latter is an inherited neurometabolic disorder affecting the central nervous system.
3. These changes are also accompanied by legislative shifts in marriage equality and the emergence of blended and LGBTIQ families.
4. While the US is notorious for private infertility treatments where IVF cycles are often very costly for the people undergoing them, countries in Europe, such as the Netherlands, Sweden, and Denmark heavily subsidize fertility treatments, opening up the practice to less wealthy couples.
5. In re Baby M, 537 A.2d 1227, 109 N.J. 396 (N.J. 1988).
6. In the UK this automatic establishment of paternity is only so in the case of unmarried surrogates. When the surrogate is married the husband is still legally seen as the automatic father of the child (whether or not genetic fatherhood is established).
7. In re Marriage of Buzzanca, 61 Cal. App. 4th 1410, 72 Cal. Rptr. 2d 280, 1998 Cal. App. LEXIS 180, 98 Cal. Daily Op. Service 1782, 77 A.L.R.5th 775, 98 Daily Journal DAR 2436 (Cal. App. 4th Dist. Mar. 10, 1998).
8. For a more thorough discussion of the reproductive bodies of postgenomics see Van Wichelen and Keaney (2022).
9. *R v. Human Fertilisation and Embryology Authority, ex parte Blood*, see also Dyer 2013.
10. *R (Mr and Mrs M) v Human Fertilisation and Embryology Authority*, [2016] EWCA Civ 611.

11. See https://www.nytimes.com/2023/04/08/style/egg-freezing-procedure-travel.html.
12. See https://www.economist.com/technology-quarterly/2023/07/17/the-fertility-sector-is-booming.
13. See https://www.forbes.com/sites/annahaines/2023/02/16/the-rise-of-fertility-tourism/?sh=66cb3fe147ef.
14. The "right to know" is defined in the Convention of the Right of the Child (CRC), which was based on the Universal Declaration of Human Rights (UDHR) and adopted by the General Assembly on 20 November 1989. Before the CRC, the "right to know" existed but was only preserved for adults. In contrast, Articles 7 and 8 of the CRC for the first time explicitly indicate that children are not excluded from this right (ibid.: 139).
15. Case of *Mennesson v. France*, Application n. 65192/11, Judgment 26 June 2014, and Case of *Labassee v. France*, Application n. 65941/11, Judgment 26 June 2014. The same judges chaired the two cases.
16. Private international law is a system of law established to harmonize and determine so-called "conflict of law" situations when private and non-state disputes cross jurisdictions. These issues usually pertain to marriage, birth, divorce, property, or commercial disputes. The Hague Conference on Private International Law (HccH) is one of the world's leading organizations for cross-border co-operation and commercial matters. It is an inter-governmental organization responsible for the regulation of private matters across different jurisdictions. Since 2010, the Permanent Bureau of the HccH has investigated issues stemming from cross-border surrogacy. Preliminary documents have been drawn up describing the information gathered from experts and/or practitioners in the field of surrogacy in HccH member states. In 2016 and 2017 expert meetings were held that included a diversity of experts from 21 States, including so-called receiving States and States of origin in relation to international surrogacy arrangements (hereafter "ISAs"). Experts included judges, governmental officials from justice and health departments, legal scholars, and bioethicists.

References

Almeling, R. 2011. *Sex Cells: The Medical Market for Eggs and Sperm.* University of California Press.

Bernstein, S. and Wiesemann, C. 2014. Should Postponing Motherhood Via "Social Freezing" Be Legally Banned? An Ethical Analysis. *Laws*, 3(2), pp. 282–300.

Besson, S. 2007. Enforcing the Child's Right to Know Her Origins: Contrasting Approaches Under the Convention on the Rights of the Child and the European Convention on Human Rights. *International Journal of Law, Policy and the Family*, 21, pp. 137–159.

Bowcott, O. 2016. Couple win legal battle against ruling on dead daughter's eggs, The Guardian, 30 June 2016, https://www.theguardian.com/uk-news/2016/jun/30/couple-win-leave-to-appeal-against-ruling-on-dead-daughters eggs

Cahn, Naomi R. 2009. *Test Tube Families: Why the Fertility Market Needs Legal Regulation.* New York: NYU Press.

Carsten, Janet, ed. 2000. *Cultures of Relatedness: New Approaches to the Study of Kinship.* Cambridge: Cambridge University Press.

Cloatre, E. 2018. Law and ANT (and Its Kin): Possibilities, Challenges, and Ways Forward. *Journal of Law and Society*, 45(4), pp. 646–663.

Dearle, M., 2015. Time for International Surrogacy Regulation: The Complex Minefield of Global Surrogacy Law Is Catching People Out and the Courts Are Left to Pick up the Pieces. *The Times*. 18 June 2015, https://www.thetimes.co.uk/article/time-for-international-surrogacy-regulation-b60rsts5ktm.

D'Alton-Harrison, R. 2014. Mater Semper Incertus Est: Who's Your Mummy? *Medical Law Review*, 22(3), pp. 357–383.

Diver, A., 2014. *A Law of Blood-ties-the 'Right' to Access Genetic Ancestry.* Springer International Publishing.

Franklin Sarah. 2013. *Biological Relatives: IVF, Stem Cells and the Future of Kinship.* Durham: Duke University Press.

Gallagher, J. 2015. MPs Say Yes to Three-Person Babies. *BBC News*, February 3. https://www.bbc.com/news/health-31069173

Gibbon, S., 2002. Re-examining Geneticization: Family Trees in Breast Cancer Genetics. *Science as Culture*, 11(4), pp. 429–457.

Haraway, Donna J. 2016. *Staying with the trouble: Making Kin in the Chthulucene*. Durham: Duke University Press.
Hird, M. J. 2007. The Corporeal Generosity of Maternity. *Body & Society*, 13(1), pp. 1–20.
Howell, S. 2003. Kinning: The Creation of Life Trajectories in Transnational Adoptive Families. *Journal of the Royal Anthropological Institute*, 9(3), pp. 465–484.
Howell, S. 2009. Adoption of the Unrelated Child: Some Challenges to the Anthropological Study of Kinship. *Annual Review of Anthropology*, 38, pp. 149–166.
Jackson, E. 2018. The Ambiguities of 'Social' Egg Freezing and the Challenges of Informed Consent. *BioSocieties*, 13(1), pp. 21–40.
Jasanoff, S. 2001. Ordering Life: Law and the Normalization of Biotechnology. *Politeia*, 62, pp. 34–50.
Jasanoff, S. 2006. Just Evidence: The Limits of Science in the Legal Process. *The Journal of Law, Medicine & Ethics*, 34(2), pp. 328–341.
Keaney, J., 2022. The racializing womb: Surrogacy and epigenetic kinship. *Science, Technology, & Human Values, 47*(6), pp. 1157–1179.
Kirksey, E., ed. 2014. *The Multispecies Salon*. Durham: Duke University Press.
Kroløkke, C. H. and Adrian, S. W. 2013. Sperm on Ice: Fatherhood and Life After Death. *Australian Feminist Studies*, 28(77), pp. 263–278.
Lafuente-Funes, S. 2023. The Role of Vitrification in Spanish Reproductive Labs: A Cryo-revolution Led by Strategic Freezing. *Science, Technology, & Human Values*, 48(4), pp. 752–776.
Landecker, H. 2005. Living Differently in Time: Plasticity, Temporality and Cellular Biotechnologies. *Culture Machine* 7. http://www.culturemachine.net/index.php/cm/article/view/26/33.
Lebner, A. 2000. Genetic "Mysteries" and International Adoption: The Cultural Impact of Biomedical Technologies on the Adoptive Family Experience, *Family Relations*, 49(4), pp. 371–377.
Lockwood, G. M. 2011. Social EGG Freezing: The Prospect of Reproductive 'Immortality' or a Dangerous Delusion? *Reproductive Biomedicine Online*, 23(3), pp. 334–340.
Melhuus, M. 2009. Conflicting Notions of Continuity and Belonging: Assisted Reproduction, Law, and Practices in Norway. *Social Analysis*, 53(3), pp. 148–162.
Mitra, S., Schicktanz, S., and Patel, T. 2018. Cross-cultural comparisons on surrogacy and egg donation. *Cross-Cultural Comparisons on Surrogacy and Egg Donation.*

Modell, J. 2002. *A Sealed and Secret Kinship: The Culture of Policies and Practices in American Adoption.* New York: Berghahn Books.

Mutcherson, K. M. 2016. Procreative Rights in a Postcoital World. *The Oxford Handbook of Reproductive Ethics*, pp. 159–181. New York: Oxford University Press.

Nash, Catherine. 2002. Genealogical Identities. *Environment and Planning D: Society and Space*, 20(1), pp. 27–52.

Nash, Catherine. 2004. Genetic Kinship. *Cultural Studies*, 18(1), pp. 1–33.

Purvis, D.E., 2012. Intended Parents and the Problem of Perspective. *Yale JL & Feminism*, 24, p. 210.

Rabinow, P. 1998. Artificiality and enlightenment: from sociobiology to biosociality. In *The ethics of biotechnology* (pp. 101–122). Routledge.

Rao, R., 1995. *Assisted reproductive technology and the threat to the traditional family*. 47, p. 951.

Richardson, S. 2015. Maternal Bodies in the Postgenomic Order: Gender and the Explanatory Landscape of Epigenetics. In Richardson, Sarah S. and Stevens, H. (eds.), *Postgenomics: Perspectives on Biology After the Genome*, pp. 210–231. Durham: Duke University Press.

Riles, Annelise. 2008. The Anti-network: Private Global Governance, Legal Knowledge, and the Legitimacy of the State. *The American Journal of Comparative Law*, 56(3), 605–630.

Roberts, D. E. 1995. Social Justice, Procreative Liberty, and the Limits of Liberal and the Limits of Liberal Theory: Robertson's Children of Choice. *Law & Social Inquiry*, 20(4), pp. 1005–1021.

Robertson, John A. 1995. Liberalism and the Limits of Procreative Liberty: A Response to My Critics. *Washington and Lee Law Review*, 52, p. 233.

Schiff, A. R. 1994. Solomonic Decisions in Egg Donation: Unscrambling the Conundrum of Legal Maternity. *Iowa Law Review*, 80, p. 265.

Shildrick, M., 2022. Maternal–Fetal Microchimerism and Genetic Origins: Some Socio-legal Implications. *Science, Technology, & Human Values, 47*(6), pp. 1231–1252.

Strathern, M. 1996. Enabling identity? Biology, choice and the new reproductive technologies. In Hall, S. and Du Gay, P. *Questions of cultural identity*, PP. 37–45, London: Sage

Strathern, M. 1999. *Property, Substance, and Effect: Anthropological Essays on Persons and Things*. London: Athlone Press.

Strathern, M. 2005. *Kinship, Law and the Unexpected: Relatives Are Always a Surprise*. Cambridge: Cambridge University Press.

Teubner, G. 2012. *Constitutional Fragments: Societal Constitutionalism and Globalization*. Oxford: Oxford University Press.
Thompson, C. 2009. Kinship in an Infertility Clinic. *Feminist Anthropology: A Reader*, p. 271. Malden: Wiley-Blackwell.
Trimmings, K. and Beaumont, P. 2011. International Surrogacy Arrangements: An Urgent Need for Legal Regulation at International Level. *Journal of Private International Law*, 7, pp. 627–647.
Van Beers, B. C. 2018. The Best Me I Can Possibly Be: Legal Subjectivity, Self-Authorship and Wrongful Life Actions in an Age of 'Genomic Torts'. In *Personalised Medicine, Individual Choice and the Common Good*, pp. 200–225. Cambridge: Cambridge University Press.
Van Beers, B., Corrias, L., and Werner, W. G., eds. 2014. *Humanity Across International Law and Biolaw*. Cambridge: Cambridge University Press.
Van de Wiel, L. 2015. Frozen in Anticipation: Eggs for Later. *Women's Studies International Forum*, 53, pp. 119–128.
Van de Wiel, L. 2020. *Freezing Fertility: Oocyte Cryopreservation and the Gender Politics of Aging*. New York: New York University Press.
Van Wichelen, S. 2016. Postgenomics and Biolegitimacy: Legitimation Work in Transnational Surrogacy. *Australian Feminist Studies*, 31(88), pp. 172–186.
Van Wichelen, S. 2019a. *Legitimating Life: Adoption in the Age of Globalization and Biotechnology*. New Brunswick: Rutgers University Press.
Van Wichelen, S. 2019b. Law's Relations in the Regulation of Cross-Border Surrogacy. In Mackie, Vera, Marks, Nicola, and Ferber, Sarah (eds.), *The Reproductive Industry: Intimate Experiences and Global Processes*. Lanham: Lexington Books.
van Wichelen, S. 2022. Identity in postgenomic times: Epigenetic knowledge and the pursuit of biological origins. *Science, Technology, & Human Values*, 47(6), pp. 1131–1156.
Van Wichelen, S. and Keaney, J. 2022. The Reproductive Bodies of Postgenomics. *Science, Technology, & Human Values*, 47(6), 1111–1130.
Vora, K. 2015. Re-imagining Reproduction: Unsettling Metaphors in the History of Imperial Science and Commercial Surrogacy in India. *Somatechnics*, 5(1), pp. 88–103.
Waldby, C. 2015. 'Banking Time': Egg Freezing and the Negotiation of Future Fertility. *Culture, Health & Sexuality*, 17(4), pp. 470–482.
Watt, H. M. 2011. Private International Law Beyond the Schism. *Transnational Legal Theory*, 2(3), pp. 347–428.
Watt, H. M. 2016. Conflicts of Laws Unbounded: The Case for a Legal-Pluralist Revival. *Transnational Legal Theory*, 7(3), pp. 313–353.

Whittaker, A. 2018. *International surrogacy as disruptive industry in Southeast Asia*. Rutgers University Press.
Yngvesson, B. 2010. *Belonging in an Adopted World: Race, Identity, and Transnational Adoption*. Chicago: University of Chicago Press.
Zafran, R. 2007. Dying to be a Father: Legal Paternity in Cases of Posthumous Conception. *Houston Journal of Health Law & Policy*, 8, p. 47.

7

Bodies: Community and Identity

> [H]ow was it possible for entire generations of human beings to reduce other human beings to the status of things, except by making the latter's bodies wholly subservient to their will? … To the process of the depersonalization of persons there corresponds that of the derealization of things.
>
> Roberto Esposito, *Persons and Things*

Introduction

In October 2018, in the run up to announcing her 2020 presidential campaign, Senator Elizabeth Warren revealed DNA results claiming to document her Native American heritage. The ensuing public discussion exposed core issues around identity, belonging, and rights in the US, a settler-colonial nation still grappling with its history of native genocide. As the anthropologists Elizabeth Watt and Emma Kowal rightly observe, the reception of DNA testing to establish communal affiliation depends on the particularity of national histories, especially histories of colonization and the treatment of indigenous groups in both the past and present

S. van Wichelen and M. de Leeuw, *Biolegality*, Biolegalities,
https://doi.org/10.1007/978-981-99-8749-8_6

(Watt and Kowal 2018). For instance, while Native Americans are suspicious of DNA testing and hostile to people making claims to Native Americanness based on such biological grounds, Aboriginal communities in Australia are less antagonistic and experience DNA testing as a "right" rather than an assault (ibid.).[1]

We start with this example to highlight the contemporary use of DNA testing as evidence of communal and national belonging. This practice is increasingly common but what does it reveal about the boundaries of community? How does biology—in this case genomic insights—contribute to the making of social identities? And how do new biological practices intervene in legal understandings of community that rely on older biological and cultural understandings of social identity? As we articulated in chapter one, the relationship between biology and community has a long history in legal and political theory, stemming from nineteenth-century ideas in *völkerkunde* (much embraced by the disciplines of anthropology and sociology at the time) and racialized practices of nation-state formation in the twentieth century. Legal scholarship has had a tremendous influence on disseminating and enforcing these supposedly scientific ideas of race, ethnicity, blood, and populations, ranging from racial chattel slavery and Native displacement and genocide, to forced sterilization, immigration quotas, and antisemitic laws in Nazi Germany (see pp. 3–4 in Whitmarsh and Jones 2010, and pp. 37–38 in TallBear 2013). Examining new entanglements of biology and community within this broader historical and temporal context allows us to understand the persistent use of biology in defining the legal contours of belonging.

This chapter investigates how contemporary biosciences reconfigure socio-legal understandings of population and community, and how the molecularization of the body and ancestry produce new forms of rights and identity. We do this by studying the forensic identification of marginal, and often criminalized, populations, to notions of biological vulnerability that internalize race, and conclude with the (potentially) inclusive remaking of identity and rights through bioscience. In all of these investigations, race and marginalization continue to constitute ideas of national belonging and identity as well as citizenship and rights. Consequently, we argue, biology contributes to determining one's (legal) belonging, with likely tremendous impacts on one's sense of community, civic identity, and belonging to the nation-state. Taking these developments together, we identify a tension between what is defined or understood as population on the one hand, and community on the other. In our final section we explore this tension, arguing that while population is the preferred bioscientific term for observation and categorization, and community is the preferred political and civic term for social groups, legal cases actively conflate biological and cultural knowledge of community around the locus of the body. Following the political theorist Roberto Esposito, we attempt to problematize law's determination of rights and identity as attached to the bounded individual.

Suspicious Populations: Molecularization and the Forensic Body

The constitutive dynamic between the biopolitical production of "populations" and its application in law is most evident in forensics and immigration. In forensics, suspects are drawn from databases that collect biological information, thus implementing biolegal actions for police and state security. At the same time, such practices challenge both personal and communal rights of bodily integrity. The forensic use of standardized "control populations" to produce suspect DNA profiles inverts the personal and collective "right to stay silent" into the always speaking "silent witness" of DNA (M'charek 2008). Such developments reveal the intimate relation between "Western cultural notions of biology and relatedness on the one hand and the codification of these categories in both law and science on the other" (see p. 269 in Hamilton 2012). The enormous increase in population databases—synthesizing human DNA with digital data—makes law the prime location for the biolegal formation and contestation of entire communities and identities.

While early successes using genetic testing to exonerate wrongly convicted individuals in the 1990s—exemplified by the well-known "innocence project" in the US—helped mainstream its use in forensic investigations, closer critical study of forensic practice reveals the new problems engendered by such technologies. Biologist and social commentator Richard Lewontin, a main protagonist in this debate, warns, as he has since the inception of DNA evidence, that when improperly analyzed, genetic evidence easily produces false convictions (Hacking 2006). The narrow demographic distribution—especially the overrepresentation of radicalized populations—fosters discriminatory practices. Like conventional fingerprinting, the size and diversity of DNA fingerprinting limits the accuracy of matches and probabilities in relation to ethnic and racial profiling because of data collection methods.[2] As it stands, most of the (national) DNA databases used in forensics primarily contain DNA collected from suspected and convicted criminals. This often means overrepresentation of minorities in the criminal justice system is reproduced in such databases, thereby limiting possible matches to this pool of people.

Other technologies have emerged alongside DNA fingerprinting that go beyond comparing individual DNA profiles. Instead, so-called Ancestry Informational Markers (AIMs) are used to gather what is now known as Externally Visible Characteristics (EVCs) of an unknown individual. AIMs involve non-gender chromosomes called autosomes. Irrespective of findings by critical criminologists and statisticians, who argue that there are systemic issues undermining the reliability of "exact matches" using this method, police now routinely use AIMs as racial and ethnic markers in forensic DNA. According to the forensic practice, AIMs determine geographical descent, allowing the reconstruction of suspects' faces—probable hair, iris of the eye, skin color, etc.—based on DNA material from other criminal investigations. Critical scholars (M'charek 2008; M'charek et al. 2020; Sankar 2010) point out that in these new DNA genotyping tools, the population definitions used are fraught with technical limitations. As predictive genetics use questionable population definitions to code the data, the data set itself is drawn from a limited population that itself reproduces the underlying problems of racial categorization and social inequalities. Moreover, against assumptions that EVCs would aid identification, these population definitions are connected with aggregate data that "provide no more than statistical probabilities calculated on the basis of aggregate data: they do not and cannot, individualize" (M'Charek et al. 2012). The way populations are defined should be transparent and contested in order to expose the intended and unintended implications of these definitions. An "unintended" consequence of technologies such as EVCs—and genetic technologies in general—is the targeting of whole communities as suspect.[3] The ethical questions these techniques raise, therefore, not only pertain to individual rights but to group civil rights, highlighting the entanglement of genetics with longstanding practices of racial profiling and stereotyping.

The making of "suspect populations" is not limited to domestic policing, as scholarly investigation into immigration practices and politics shows. For instance, in Europe and beyond, biological evidence in the form of kinship testing is increasingly required of families seeking reunification (Heinemann et al. 2015). The uses of biometrics, including finger and facial scans, are now also routine features of border security and the verification of identity. Its widespread application in immigration normalizes the datafication of the migrant body, constituting it as an object for data collection and surveillance (Pugliese 2012). Border services in Canada for instance, have submitted migrant DNA to direct-to-consumer genetic testing companies to determine detained migrants' nationalities, despite the fact that ancestry testing only traces people back to "ethnic regions" or "populations," and not nationality, which is a legal category (see p. 3 in Abel 2018). These technologies are used freely regardless of the obvious ethical concerns they raise—not least of all the issue of informed consent for refugees.

More complex genomic sequencing and epigenetic research have complemented forensic and immigration practice. For instance, border enforcement currently uses the so-called "epigenetic clock," which measures methylation across the genome, to authenticate the age minors provide on their asylum applications (Ihar 2020). These developments further the molecularization of the body, as well as of race and ethnicity. All these technologies prioritize the body as the location of truth by mobilizing genomic data to evaluate and substantiate historical or anecdotal documents (see p. 342 in Benjamin 2009: 342, and Fassin and D'Halluin 2005). As scholars have pointed out, the preoccupation with population genetics for forensic and immigration uses strongly contributes to the idea that evidence and truth are distilled and objectified in the body (Fassin 2013; Heinemann and Lemke 2014). As such, documentary evidence, while still the primary evidentiary source in many jurisdictions, competes for recognition against newly established biological evidence taken from the body. In doing so, these technologies appear to substantiate the idea of suspect "populations."

While scientists have recently turned to populations to replace previous categories of race, critical scholarship demonstrates that uses of populations, especially in contemporary forensics and immigration policy, collapse into existing categorizations (i.e. race) based on uneven data collection. As noted earlier, genetic databases are often incomplete and irregular, with certain populations frequently overrepresented, producing suspicion of these (often racialized) groups. Moreover, the introduction of ethnicity and nationality into population delineations further complicates the application of scientific evidence in forensics and border control. Some of these applications have been brought to courts and emerge in the jurisprudence around DNA testing.

Use of DNA testing in courts has been discussed at length, particularly concerning paternity testing and the right to access DNA testing when making citizenship claims in family reunion cases (Davis 1994). Sociological investigations, however, give strong accounts of the politics underpinning the biolegality of genomic and postgenomic technologies, exposing the assumptions underlying compulsory DNA testing and pointing to legal problems of consent, privacy, and human dignity (Dove 2013; Heinemann and Lemke 2014). These scholars suggest that while DNA testing can help families reunite, DNA testing in some jurisdictions' forensic and immigration practices—especially when they're compulsory—sustains older and racialized logics of exclusion, where the "other" is always already suspect and in need of scrutiny.

Geographical Populations: Difference, Ancestry, and Community

The development of genomic human populations studies, which, using ancestry, maps genetic variances "onto traditional racial categories," granted greater validity to the rise of DNA profiling in forensics (Whitmarsh and Jones 2010: 6, see also Bolnick et al. 2007; Dupré 2008;

Nelson 2008). Starting in the 1990s, the Human Genome Diversity Project (HGDP) set out to sample, map, and archive human genetic variation from populations around the world. Although framed as an independent enterprise, the HGDP is an important complement to the Human Genome Project (HGP). The full sequencing of the human genome revealed that 99.9% of human DNA was universally shared, conveying the message that "long-term antiracist efforts to de-biologize race were legitimized by scientific findings" (Hamilton 2008). In contrast, genetic diversity research (including HGDP, but also the International Human Haplotype Map or HapMap and the Genographic Project) focused on the 0,1% of human genetic variation to advance knowledge about genetic geography, history, and evolution. Although promoted as cosmopolitan projects, they reignited thorny debates around race and ethnicity (though now discussed in terms of population).

Central to the HGDP was the necessity to collect and analyze the genetic profiles of indigenous populations because their lack of "admixture"—relatively endogenous populations with homogenic ancestry—was regarded as a reliable baseline for tracking the migration of people in the past. Lead by the population geneticist Luigi Cavalli-Sforza, the HGDP published their findings of the whole genome sequencing of indigenous populations in 2020, accounting for the genetic variation of fifty-one populations across Africa, Asia, Native America, Europe, and Oceania.

Called the Vampire project by indigenous rights activists, and biocolonialism by anthropologists, the HGDP raised significant social concerns. Besides the ethical issues around informed consent, critics warned that the study of difference in terms of genetic makeup carried the potential for discrimination. More pointedly, however, critics challenged the appropriation of indigenous DNA for scientific purposes. The Genographic Project, a privately funded outgrowth of the HGDP, was seen as illustrating such appropriation. The project focused on tracing the history of human migration around the world by collecting and analyzing indigenous DNA. The DNA stories coming out of this investigation, however, clashed with the epistemic knowledge of indigenous communities, in some instances jeopardizing indigenous claims to land and sovereignty (Reardon and TallBear 2012; TallBear 2013).

Throughout the HGDP and the Genographic Project, social scientists called for more ethical procedures to guide interactions between scientists and indigenous research subjects. They also challenged the methodological presuppositions, pointing to how settler-colonial tropes—where "whiteness as property" (Harris 1993) legitimated the removal of indigenous lands—were transferred to contemporary practices in bioscience (see p. 235 in Reardon and TallBear 2012).

Designed in 2002 to counteract the social concerns raised by the HGDP (see p. 472 in Hamilton 2008), the HapMap was international, involving public and private research centers from Canada, China, Hong Kong, Japan, Nigeria, the UK, and the US, and through final publications came to completion in 2010. The HapMap study examined the genetic causes of chronic non-communicable diseases. Rather than focusing on the "lack of admixture," as was the case with the HGDP, mapping admixture was the goal of the HapMap, with the intent of identifying particular groups of people to study disease risks. Specifically, the HapMap was explicitly concerned with genetic difference as expressed by single nucleotide polymorphisms (SNPs). These SNP markers denoted "population-specific alleles," later categorized as "ancestry informative markers" (AIMs), which, as discussed in the previous section, were put to forensic use. The project's underpinning assumption is that "genetic variability exists among large-scale populations — broadly defined as African, European, and Asian — and that this variability is linked to human health and disease in significant ways" (see p. 473 in Hamilton 2008). Unlike the HGDP, though, ethical, legal, social implications (ELSI) were central to HapMap's design. The sociologist Jenny Reardon describes these inclusions as representing the rise of "genomic liberalism": the active formation of participatory, open, and inclusive genomics (Reardon 2017). Rather than limit the power and control of research to scientists and their funders, genomics would be democratized, extending beyond the field of researchers and practiced by all people.

In contrast to the HGDP, it was not difficult for HapMap to enroll "communities" in their project. On the contrary, participant groups were reportedly honored by their involvement in such a prestigious scientific project and offended when they were overlooked (see p. 74 in Reardon 2017). Reardon explains, such dynamics provoked bioconstitutional mechanisms, enabling people and communities to develop new rights in relation to their genomic identities, becoming constitutive members and citizens, through enrollment, of genomic initiatives.[4] The early stages of the HapMap, then, saw encouraging movements toward the integration of ELSI practices via policies requiring community engagement. Yet, as several scholars point out (Chow-White 2012; Hamilton 2012; Rajagopalan et al. 2017; Reardon 2017), "community" lost meaning as it traveled outside of the US borders.[5] While care was taken to talk about populations, the boundaries of African, Asian, and European ancestry were increasingly conflated with the US racial taxonomy and therefore became increasingly contentious (Chow-White 2012: 92). Moreover, as Reardon explains, the ELSI policies "did not address the bioconstitutional questions that faced the HapMap: Who are the people who should have the right to be included and represented in maps of the human genome?" (see p. 92 in Reardon 2017).

In all these initiatives, the state did not idly stand by. Iceland, the UK, Scotland, Canada, Taiwan, Singapore, and Brazil launched national biobank projects during this same period, giving rise to what has been termed "biosovereignty" (see p. 40 in Ong 2010, and pp. 60–62 in Ong 2016) or "genomic sovereignty" (Benjamin 2009). Here processes of making value and meaning are tethered to biomaterials in order to secure new economic assets in the growing bioeconomy. Acting as biolegal tactics, states limit global market forces to protect their biological resources (see also Van Wichelen 2023). Increasingly, then, biology, and in particular biomedicine, is wrapped up in the practices and rituals of citizenship—whether this pertains to the state's attitudes toward citizens' health, or to citizens' concerns regarding access to and quality of health care (see p. 9 in Whitmarsh and Jones 2010).

Away from national or global scientific projects such as the HGDP and the HapMap, the use of ancestry markers was also offered directly to consumers (DTC); on the one hand, tapping into what came to be known as personalized or precision medicine (discussed in the next section) and, on the other, supposedly providing scientific assurance of one's ancestral history. But communities respond differently to the introduction of these genetic tests. The anthropologist Kim TallBear describes, for instance, how DNA stories disrupt Native American legitimating frameworks attached to blood ancestry. DNA testing companies in the US increasingly claim they can link ancestry markers to specific tribes and promote their services to assist Native Americans in furthering their rights to certain benefits (see pp. 85–86 in TallBear 2013). But tribal leaders have resisted membership based on DNA testing. Instead, blood regimes—based on historical and social practices of blood lineage—run strong in tribal communities and are decisive in the recognition of true affiliation.[6]

While Native American communities have met genetic testing with animosity, or at least ambivalence, African American communities have embraced this new technology in their pursuit of justice and reconciliation. As the sociologist Alondra Nelson so vividly describes in her work on ancestry testing among African American communities in the US, practices of DTC testing "allow us to try—or try again—to contemplate, respond to, and resolve enduring social wounds" (see p. 7 in 2016). Black root seekers were interpellated by the explosion of media attention to genetic testing, epitomized by the hugely successful public television program *Finding Your Roots* by Henry Louis Gates Jr., as well as by the scientific and cultural guidance of major African American scientific entrepreneurs such as Rick A. Kittles, founder of African Ancestry Testing. Nelson's point is that the meaning of genetic testing and knowledge for African Americans goes beyond personal revelation and is embroiled with the making of diasporic communities in the wake of natal alienation during the chattel slave trade, and part of a continued commitment to racial reconciliation and justice. In this case, the social life of DNA constituted communities and with it renewed civil calls to legislation on reparation—a much avoided topic in US history[7]—now with biosocial grounding. More generally, current practices of ancestry testing indicate, as Faubion and Hamilton argue, that the biogenetic understandings of kin, clan, tribe, and community undergirding these practices are deployed and used both by institutions and citizens or citizen-groups to reshape political and legal affiliation, (see p. 276 in 2007). This tension reverberates in discussions on the rise and promise of precision medicine.

Personalizing Populations: The Rise of Precision Medicine

The so-called "omics" revolution in the biosciences is changing medicine and its practice in corresponding health institutions. Following the human genome project, and accomplishments in genomic sequencing, other "omics" have emerged—including proteomics, metabolomics, and epigenomics—using vast amounts of data to model or make predictions computationally and statistically based correlations with health outcomes (O'Malley and Skillings 2018; Stern 2019). The intensification of genomic sequencing in research, precipitated by technologies making it easier, quicker, and more affordable to sequence whole genomes, led to the new field of "precision medicine," more commonly branded as "personalized medicine," and in critical contexts, also referred to as "race-based medicine."[8] Precision medicine promises to find new ways of treating patients based on personal biological markers. These biomarkers undergo genomic analysis, referred to as single nucleotide polymorphisms (SNPs), to deliver risk calculations (proportionate to the genomically categorized population) for certain diseases or conditions. Hailing "precision medicine" as the most promising development in contemporary healthcare in 2016, President Obama and the US state invested $215 million toward its expansion, with Europe, Australia, and other regions following suit. Asia too is heavily invested in this initiative and is often labeled the new frontier for postgenomic medicine (Ozdemir et al. 2011; Sun 2017).[9] While precision medicine's potential cancer therapies are especially highly touted, other "global" and chronic diseases, such as diabetes and Alzheimer's, are targets of multi-national research funded by such initiatives.

Pharmacogenomics is the most prominent branch of precision medicine, with drug therapies catered to individual (genomic) characteristics to minimize adverse reactions. The detection of biomarkers via vast computational techniques goes beyond genomics and includes other "omics," such as proteomics (the study of proteins), metabolomics (the study of chemical processes involving metabolites), and epigenomics (the study of epigenetic modifications). Proclaimed the so-called "omics revolution" by media, these studies (like genomics) use new computational methods to process massive quantities of data to find correlations that associate medical conditions to certain biomarkers. The growth of biobanks across the world muddles the legal and regulatory dimensions of precision medicine, something data-sharing across multiple jurisdictions amplifies.

Major changes to medical research and health care exacerbate already complicated issues of consent, privacy, and confidentiality with, for instance, the "blurring of boundaries between the clinic, the laboratory, and the healthcare industry, creating new regulatory spaces" (see p. 284 in Nicol et al. 2016). While in the past clinics were the only institutions able to deliver and manage health-related services, private labs, and for-profit enterprises, conducting research for pharmaceutical companies or performing Direct-to-Consumer (DTC) testing, define precision medicine's spectacular rise and ambitious promises, and extend clinical practice beyond the clinic.

The privatization of genomic research and the flourishing of the DTC genetic testing market, exemplified by the US company 23andme, not only changed the landscape of health and research institutions—and their attendant legalities—but also changed our somatic experience and negotiation of health risks (Novas and Rose, 2000; Rose 2007). Timmermans and Buchbinder argue that the health risks articulated in genomic screenings—often imprecise or unintelligible to the lay person—confront patients with diagnostic uncertainty, creating an experience of "prolonged liminality between a state of normal health and pathology" (see p. 408 in 2010). Some argue this situation can be improved by providing genomic education—or soft regulation—to assist people in navigating and interpreting the biological information they receive (with the help of, for instance, genetic counselors). Nonetheless, with the discovery of consistent correlations between biomarkers and certain health conditions, one can predict, with Timmermans and Buchbinder, that this new pool of "patients-in-waiting" will increase as the growing availability of personalized genomic information continues to produce new forms of biosociality constituted around the body as always already potentially diseased.

Regardless of precision medicine's potential for individual health and patient empowerment, its increasing pervasiveness, as critical science scholars point out, carries pitfalls and risks both within and outside of biomedicine. Important for our argument here is the central role populations play—as they do in forensics—in precision medicine. Although the terms "personalized medicine" and "precision medicine" assume that diagnosis is made on the level of the individual, the science underpinning these practices relies on population level data in the aggregate. Precision medicine promises to assess individual risk but consists of comparing groups of people (see p. 74 in Hinterberger 2012; see also Green et al. 2019; Jones 2013). Rather than the individualization of medicine, Dickenson et al. (see p. 3 in 2018) argue it is more apt to talk about the stratification of medicine at best. Biomedical scientists stratify patient groups according to ancestry and ethnicity, and subphenotype (individual) patients according to these groups to see how they respond to certain drugs (ibid.). Precision medicine's promise, and its neoliberal narratives of patient empowerment and individual choice, hides the practice of stratifying groups of people to identify biomarkers.

A considerable part of precision medicine is practiced through genome-wide association studies (GWAS), which provide evidence of correlations allowing for targeted treatments, primarily of Crohn's disease, schizophrenia, type 2 diabetes, and cardiovascular disease. However these studies, have been scrutinized for their heavy reliance on the accumulation of health data—merging with the vagaries of the unstructured world of Big Data—raising ethical and legal concerns with respect to privacy, ownership, and the status of the data being analyzed (see p. 579 in Chalmers et al. 2013, see also Prainsack 2017; Hoeyer 2019; Dickenson 2013; Van Beers et al. 2018). As Hinterberger argues, even though molecular biology is less deterministic than in the past, GWAS "still must pass through a series of scales and grids (molecular, national, and global, to name a few) that constrain, define and limit their promises" (see p. 80 in 2012). Moreover, while GWAS researchers take care not to ascribe genetic similarities to certain populations or races, the reception of early studies revealed a sustained preoccupation with race as having a biological component (Fujimura and Rajagopalan 2011).

Further to suppression of stratification processes in order to popularize the more amenable discourse of personalization in the public domain, precision medicine does not acknowledge the social context of data, even though the social deeply informs the biological in their genomic studies. As Hinterberger shows in her ethnographic study of a private Canadian genomic lab, GWAS' use of, for instance, "aspects of clinical work, sample collection, national census statistics and political doctrines such as multiculturalism" brings the social into the biological (here the SNPs), in turn, leading to the biologization of the social. The anthropologist Klaus Hoeyer (2018) makes a similar point with regard to data "sourcing": data are not *found*—as the concept of data "mining" suggests—but are "made to exist" through a process that transforms individual data into population data, which then returns, in nearly tautological form, to the individual as diagnoses or health advice (see p. 137 in 2018). One could argue that such uneven feedback allows for what Ian Hacking calls a "looping effect" (2006) where populations become moving targets as the studies themselves change the boundaries of these categorizations.

These dynamics are especially delicate when populations correspond with ethnic or racialized groups, thereby reinvigorating older debates around human difference and race (Bliss 2012; Duster 2006; Nash 2005; Roberts 2008; Whitmarsh and Jones 2010). Critical scholars highlight this racial dimension by referring more explicitly to race-based medicine rather than precision medicine or singling out studies in precision medicine that specifically address race. Aimed at connecting genomic predispositions to racial health disparities (for instance cancer or heart disease), race-based medicine contends that more targeted therapies can then be developed based on these racial categorizations. A prominent example in the US is the NitroMed, Inc. drug BiDil, which is marketed specifically for African Americans suffering from congestive heart failure. Within research communities in the life and the social sciences, the distribution of BiDil revived the debate as to whether there was a genetic basis for race (Fullwiley 2007; Dorr and Jones 2008; Pollock 2008; Whitmarsh 2009; Roberts 2008, 2011). Arguing against linking race and genetics, scholars critique the reinscription of race through the molecularization of the body (Abu El-Haj 2012; Duster 2006; Ossorio and Duster 2005; Fullwiley 2007).

The problem with BiDil—and race-based medicine generally—is that genes and geographic origin (biological race) are mixed with self-identification based on historical racial categorization, thereby producing an incommensurability in the very substance of race (Duster 2006, see also Reardon 2009). At the outset BiDil was not meant to cater only to African Americans. The drug's components—a combination of hydralazine and isosorbide dinitrate—were already approved by the FDA for the treatment of hypertension and angina and prescribed to patients starting in the 1980s (Dorr and Jones 2008). It was not until the FDA rejected the application from NitroMed for an extension of their patent that the company started clinical trials based on racial correlations in a bid to revive the patent. After these analyses showed a strong association between BiDil and the health improvement of African American men, new race-specific trials were organized involving participants "self-identified as black (defined as of African descent)," which proved so successful that the study was terminated for ethical reasons around denying the control group the treatment.[10] Soon after, BiDil won approval from the FDA in 2005, marking the first time the FDA has ever approved a race-specific drug. While lauded in mainstream media as an antidote to racial disparities of health, BiDil also received substantial scholarly criticism in terms of using race-based pharmaceuticals to address inequality while excluding consideration of race from public policy decisions (see p. 1 in Whitmarsh and Jones 2010, see also Roberts 2008, 2013).

Important for our purposes here is the extent to which existing laws and regulations played a significant role in creating the conditions out of which race-based medicine evolved. The rejection of a patent application prompted the race-specific analysis of previous trials, which then became constitutive in the formation of race-based medicine. As critics have noted, this precedent is problematic on several fronts: First, the FDA approval of BiDil seemed to signify an economically motivated use of loopholes in US patent law where addressing racial health disparities enabled the extension of a previously rejected patent. Second, despite the reliance on racial self-identification (rather than genetics) in the BiDil trials, the FDA's justification for its decision and the subsequent marketing and distribution of the drug implicitly made a claim for a genetic link, therefore contributing to a new regime of race-specific pharmaceuticals (Kahn 2004).

In response to these critiques, scholars emphasize the potential of invoking Equal Protection jurisprudence and constitutional law in these matters (Kahn 2006; Obasogie 2008; Roberts 2006). For instance, legal scholar Osagie Obasogie argues for adopting of the doctrine of "strict scrutiny" common in constitutional law to negotiate a workable public policy or regulatory framework—one that acknowledges the health benefits of race-based medicine while attending to the risks of "geneticizing race" (see p. 494 in 2008). Applied to biomedicine, "strict scrutiny" of race-specific trials and drugs would assess the validity of their purpose, the robustness of research, and the intervention of markets (Winickoff and Obasogie 2008), supplementing the rigorous clinical examination of "safety" and "efficacy" with similarly rigorous testing of social responsibility (see p. 495 in Obasogie 2008). Such an approach singles out race as a unique matter within clinical bioscience, one which demands specific scrutiny given the history and present conditions of race relations.

These developments and debates in race-based medicine are precursors to future debates on precision medicine more generally. In different states and jurisdictions across the world, laws have been passed banning genetic discrimination (Joly et al. 2010). While these laws claim to offer a measure of protection for citizens-consumers, they are also a legal requirement for advancing personalized medicine as an industry. The sociologist Dorothy Roberts (2013), for instance, writes about the US Congress' 2008 Genetic Information Nondiscrimination Act (GINA) as representing a neoliberal creation of law supported by drug companies, scientists, and patient groups. Roberts notes that while the law allows the precision medicine industry to flourish, it does little to assist people who are sick or those who are subjected to health disparities in society, pointing to simultaneous policies in welfare, prison, foster care, and immigration that reinforce structural inequalities for already marginalized and racialized communities (ibid.: 271). In sum, the debates around race-based medicine—and precision medicine more generally—explore not only the uses of race or populations in research, but also signify the ways our self-conception as members of a community or collective of biocitizens is implicated (or not) in these endeavors. While the rising power of genomic knowledge necessitates legal protections to prevent a "backdoor to eugenics" (Duster 2006), law also facilitates practices and institutions that reify the biologization of race (Duster 2015; Montoya 2007; Virruel-Fuentes 2007). In the next section we describe how postgenomic sciences consolidate the idea that social experiences of racism or trauma enable disease to "become" biology.

Embedded Populations: Postgenomic Plasticity

What happens to the legal idea of community and identity when postgenomic sciences render the human body porous and plastic, rather than bounded and determined by static genetic codes? As chapter two discussed, the recent decades witnessed the emergence of a host of postgenomic sciences encompassing developments in neuroscience, epigenetics, microbiomics, and immunology. Sometimes referred to as the "new biologies," these developments challenge the more deterministic orientation of genetics evidenced by the predictive forensics and population-based biomarkers presented earlier in the chapter or the neuroscientific criminalization in chapter three, where genes, cells, and brains are fixed and bounded. Instead, postgenomics portray biological life as plastic. Plasticity here refers to the flexibility and openness of the body, as well as the constant exchange between the body's internal functions and the external environment. Here, biological processes are less linear and determinant than the sciences once thought. In its place, bioscientific insights point to complex permeabilities and malleabilities. Considering the restricted space, we limit our focus here to epigenetics, whose scientific outcomes have had considerable implications in the ethico-legal space of rights claims, including communal ones. As a form of historical genetics detailing the genetic inscription of change over time, postgenomic plasticity becomes a site for contestation over lasting harms that are both physical (cancer) and social (generational trauma).

In a speculative manner discussion has started on the implications of epigenetics for law and legal practice. Often, they are about litigation, involving claims of individual harm based on epigenetic analysis, especially regarding chemical exposure and epigenetic effects on the formation of certain cancers. Such debates mirror cases in genomic torts, where a physician or medical professional's negligence in genetic testing has caused harm to the litigating person, which results in legal liability dependent on three legal conditions: the act, the injury, and the causation (see p. 797 in Dupras et al. 2019). According to Dupras and colleagues, the causality clause proves hardest in court as judges are not keen to rely on population studies to establish evidence of causality. Epigenetics changes this and provides the missing evidence that shows causality beyond population studies (ibid.), proving beneficial on the individual and communal level. For instance, epigenetic research that give evidence of racial health disparities might be used to make claims of harm to the community as a whole. In the realm of indigenous rights, for instance, the use of epigenetics as leverage in discussions of reparations raises the stakes of political and social justice claims (Warin et al. 2020). The science of environmental epigenetics provides indigenous communities an evidentiary context, connecting the colonial and traumatic past to their present biological existence. Within the legal field such claims have the likelihood of broadening the context of "discovery of harm" to include multi-generational and intergenerational effects of historical trauma.

Here new bioscientific insights—and "postgenomic styles of thought" (see p. 391 in Meloni 2017)—affect how individuals and communities think about rights concerning identity and culture. Emerging scholarship from the field of (assisted) reproduction have, for instance, raised questions about the correspondence between epigenetic harm and social, racial, and economic disparities (Lappé et al. 2019; Guthman and Mansfield 2013). A certain style of genetic thinking now permeates Western society, redefining how we relate to identity and community, in terms of inclusion and exclusion, both socially and legally. The postgenomic field of epigenetics introduces new terms into this debate by offering a biosocial model of life (Meloni 2017; Niewöhner 2011) where social and physical environments (ranging from traumatic conditions to toxins and nutrition) rearrange biology for multiple generations.

However, deployments of the new science of life have also brought tension and contradiction around how plasticity can be understood.

Critical social scientists caution against celebratory appraisals of the bioscience pointing to limitations especially in the ethico-legal space around reproduction (Karpin 2016; Manderson 2016; Valdez 2018; Van Wichelen 2016; Van Wichelen and Keaney 2022; Vora 2015) and around race and identity (Landecker and Panofsky 2013; Mansfield 2012; Mansfield and Guthman 2015; Meloni 2015; Meloni and Testa 2014). Regarding reproduction, the womb is a key site for epigenetic research, and social science scholars have addressed how the pregnant person is mostly presented as environment for another person, thereby individualizing risk by molecularizing maternal responsibility (Lappé 2016: 680–681). With respect to race, scholars are concerned with how reductionist takes on gene environment could further racialize communities, particularly as marginalized groups are portrayed as exemplifying epigenetic damage too complex to address by public health policies (Pickersgill et al 2013; Meloni 2017). As stigma, discrimination, and social stratification leave a heritable biological trace, social disparities of race become etched in biology, offering a platform for biosocial theorizations of racism and social violence, but also opening yet another door for increased scrutiny of marginalized and racialized populations (Lock 2013; Mansfield and Guthman 2015; Niewöhner 2011).

Law Embodied

Rather than emphasize law reform to regulate emerging technologies, the social science literature detailing developments such as population genetics or precision medicine highlights the ways communities and citizens are caught up in processes of social identity-making, citizen-consumerism, and knowledge production (see p. 91 in Jasanoff 2004). Not only are ordinary and marginalized people attempting to partake in the knowledge society but, as Jasanoff argues, "just as urgently, states and state-like institutions are finding it necessary to ask how they can retain their legitimacy in an era when knowledge is distributed and decentralized" (see p. 92 in ibid., see also Nowotny et al. 2001). How boundaries of groups and collectives are drawn is at the core of the relationship between (political) communities and the state. Bioscience disrupts and alters how these boundaries are made and maintained and, in turn, the groups and collectives they define. It is therefore at once a product of the relationship between community and the state and poses challenges to it, which the law then attempts to mediate.

Indeed, the history of "sanctioned identities" (Collier et al. 1995) goes back to the construction of legal subjects via apparent "natural" differences (belonging to persons outside of legal relations) but also the rights that come with these differences to demand justice. The nineteenth century witnessed the peak of using cultural difference to delineate the nation-state and to legitimate revolutionary struggles for freedom and self-government (Anderson 1991; Hobsbawm 1992; Chatterjee 1993). A version of this is repeated during the twentieth century in Asia, Latin-America, and Africa, with anti-colonial movements fighting for independence: "While agreeing that men of reason must govern themselves, they invoked cultural traditions to delimit the 'nation' and to define the 'selves' eligible to participate in governing it" (p. 13 in Collier et al. 1995). Feminist, civil rights, queer, disability, and anti-racist movements, further refine and complicate the connection between identity and community in relationship to liberalism and its legal regimes. All are premised and dependent on both universality and particularity. Here, a dialectical and contradictory relationship exists between formal identity as empty universal (the historically unmarked and rights possessing white man) and the categorically instantiated identity of identity politics that is racialized, ethnicized, sexed, and gendered, where law seems to shift between these positions as a sort of mediator[11]:

> First, bourgeois law declares everyone equal before the law, but by doing so constructs a realm outside of law where inequality flourishes. Second, bourgeois law simultaneously demands and disclaims difference, requiring people to have unique identities and individual wills while compelling them to stress their similarities with other abstract bearers of legal rights if they wish to be treated as equals. (p. 21 in Collier et al. 1995)

It is in Euro-American modern law that we witness the formalization of sanctioned identities, where subjects are legally defined according to their natural—and later biological—capacities and needs (ibid.). Because "natural" difference supposedly precedes law, it became the foundational legal principle on which human equality before the law was based. Feminist and postcolonial theory have forcefully demonstrated that modern law's construction of the natural and universal Man reflected and reproduced patriarchal, (white) imperialist societies (Davies 2023; Harris 1993). Nonetheless, as the women's and civil rights movements across the Western world and beyond testify, modern law is elastic, expanding the formal arena where demands regarding the right to difference are recognized. Law, then, stretches itself to incorporate a normative idea of plurality, but always as attached to equal individual liberties (Habermas 1996; Young 1990). With the advancement of biotechnology these natural categories, dubiously identified as 'natural' in the first place, are now disputed biologically. How far can law stretch itself to include the formation of new communities that exceed "natural" givens and are, instead, informed by new biological allegiances?

Thus, the biotechnological developments described in this chapter signal two things in the realm of law and legality: first, they trouble the idea of naturalness at the level of biology rather than social construction; and second, they engender biosocial approaches for claiming the right to equality across difference. The uses of biology in modern law, then, contradicts and recognizes community at the same time. While underplaying biological difference by insisting on sameness (genetically, we are 99.9% the same), biolegality also affirms the specificity and singularity of biological relations (as with genetic ancestry testing or national genome projects like the HGDP or the HapMap). But on both sides of the genetic discourse, the liberal subject of law remains intact, and the liberal community remains uniquely immune to life conceived of in terms of an aggregate of individual persons, not necessarily as a social body. Put differently, within the liberal context, community can only be understood as aggregates of individual identities.

This is why identity politics and the "law of the common" pose a significant problem for the liberal framework of rights. Does working toward a law of the common mean that individual life should be claimed by the "community" as a form of identity politics? Sociologist Yasmeen Arif (2016) argues that both communal (based on difference) as well as individuated immunity (liberal) fail to care for the self. And this is reflected in Euro-American law, especially in our human rights regimes. The framework of "rights" works only through the claims of a certain formal, universal, and fixed identity. You cannot invoke law unless you have this marked identity; hence law incorporates and prescribes community as a form of immunity (Esposito 2013), broadly functioning to protect members against external but also internal threats (think terrorism and pandemics). New allegiances based on biology further disrupt the crisis of the community-subsumed-by-immunity.

The challenge today might be to think of identity not as a form of community or "identity politics" but, instead, as a provisional place that is geared toward "life politics" (Vatter and De Leeuw 2023). In relation to the law, such an affirmative biopolitical position refuses and counters the disembodiment of liberal rights discourse and its incorporation of identity politics. As the legal theorist Patrick Hanafin argues:

> Within this space of relation above the law what is performed is a right to refuse to be rendered disembodied which is so powerful because not given by some divine power or moral code. This right to refuse disembodiment calls us to think beyond the State and our role in the biopolitical matrix. It is a praxis of politics as relational. (see p. 87 in 2008)

Hanafin goes back to Agamben to illustrate how the state can recognize claims of identity but is incapable of tolerating singularities that form a community regardless of a common identity: "Whatever singularity ... [that] rejects all identity and every condition of belonging, is the principal enemy of the State" (Agamben quoted in Hanafin, see p. 87 in 2008). A relational notion of rights radically alters the legal regimes underpinning the naturalness of identity in modern law. It would be a politics of "singularities without identity, without qualities, whos not whats" (see p. 88 in Hanafin 2008). Such a praxis refuses to fix persons to one place through legal decrees and would acknowledge embodiment as part of the politics of life.

In the Western tradition, the notion of rights (including human rights) cannot escape the confines of identity. If law regulates how bodies—the body as person, baby, refugee, migrant, minority, sick, disabled, and so on—interact with each other, the changing biological definitions of—or interventions in—the body determine its legal status and vice versa. Law as a biolegal regime, then, not only decides who is a person or a thing, who owns or is owned, but also, who can form or belong to a community, who can claim minority status (as Indigenous, Asian, African American, and so on), who can access public institutions like schools (if vaccinated), or who can be a lawful citizen. Summing up, the turn to Agamben and Esposito moves away from law's commitment to formal equality by way of subsuming identity politics, reifying identity by fixing categories, and effacing how these categories were socially constituted (by naturalizing identity). An affirmative biopolitics would be organized, not around a universal liberal subject, nor around the categorically articulated, fixed, and naturalized subject—who are both legitimate possessors of rights. Instead, an affirmative proposal would be structured more readily around a community of life founded on a biosociality that considers both singularity and relationality (Hinchliffe et al. 2016; Ingold and Palsson 2013; Lock and Palsson 2016). A legal shift to "the body," rather than "the person," could more adequately clarify the stakes involved in the biolegal configurations of communities and further prepare us for our posthuman futures. Moreover, this paradigmatic shift opens up a new approach to rights, one that is more attuned to the singular, relational, and permeable body that transgresses Euro-American privileging of the person.

Conclusion

This chapter showed how advances in genomic and postgenomic technologies alter our sense of identity and community. As Jasanoff argues, genetic information can never be wholly personal: "it is also information about a person's family, clan, tribe, or ethnic community" (see p. 181 in 2016). Moreover, these changes are not detached from society, law, or the state; they affect how—as citizens and non-citizens—we reorient our notion and understanding of rights and justice. While the turn of the twenty-first century signaled the emergence of a new form of biological citizenship with increased access to health data via the internet (Rose and

Novas 2005), biological citizenship became increasingly complex and fragmented as people not only incorporated scientific knowledge about individual health into their identities, but also began adopting new ways of being and relating, while negotiating risks and probabilities of disease and disability.

As such, these new biocitizens are themselves entangled in biolegal practices, from securing their rights and entitlements in the uses of genomic and postgenomic technologies, to transgressive forms of legal identity precipitated by practices from the margin, such as postmortem or post-conflict identity recovery in forensics (Gonzalez-Polledo and Posocco 2021, see also p. 753 in M'Charek and Casartelli 2019). Many of these advancements, ranging from genomics to reproductive technologies, neuroscience, and precision medicine, catalyzed processes of *subject-making* (Hamilton 2012: 268; Van Beers 2018). In turn, these developments produce biolegal shifts in our understanding of community.

While emerging technologies gain private purchase in the global bioeconomy—exemplified in this chapter by the appropriation of indigenous DNA, the rise of DTC genetic testing, and the manipulations of patent law around race-based medicine—the state remains a central locus in determining the contours of biological citizenship. As states utilize technologies that rely on the study of populations—in forensic and immigration practice, health and prenatal screenings, and the facilitation and regulation of commercialized testing—they also continue to use biology and biolegality to govern populations and communities. This goes beyond the idea of disciplinary governance or the control of biosecurity; following Reardon's proposal for genetic liberalism, progressive ideas of freedom and open information are etched in the project of biomedicine, affecting how we as communities and societies give meaning to life and law.

In our final reflection, however, we tried to think beyond the contemporary framework and its diagnosis to reimagine identity and community in a new framework of rights. A legal shift to "the body" rather than "the person" more adequately clarifies the stakes involved in the biolegal configurations of communities and better prepares us for our posthuman futures where not only fellow humans, but other species inhabit our imagination of collectivity. Such a paradigmatic shift could also formulate a new approach to rights, one that is more attuned to the communal, relational, and permeable body, transgressing Euro-America's privileging of the person.

Notes

1. As Watt and Kowal argue, the Australian viewpoint can be explained through the specific history of the "stolen generation," which led to many aboriginal descendants not being able to trace their ancestry via communal means, nor were they able to live "black" lives because they were raised in white families. Their tragic fate made it possible for aboriginal communities to accept claims of affiliation based on DNA testing.
2. The databank of DNA fingerprinting looks much like the hollerith card bank of fingerprints in the past but instead of large buildings physically holding cards containing fingerprints, the system is automated and electronic (though not less vulnerable for human error).
3. We bracketed the word "unintended" because of course within a context of institutionalized racism, one can see the practice as serving the purpose of reproducing historical racial hierarchies and therefore not unintended. Thanks to Ben Stork for making this point visible.
4. Reardon refers here to Jasanoff's term "bioconstitution" (2011), which we discuss in our introduction.
5. As Reardon notes: "The HapMap 'community' in Japan sought to represent 'Asia.' Another in southern Europe sought to make visible the history of Western invasions. HapMap leadership wanted to avoid 'confusion,' and so embraced 'precision,' which they defined as using specific geographic locales to represent those they sampled […]" (see p. 85 in Reardon 2017).
6. In US history, tribal belonging was core to the construction of limited sovereignty and negotiations over land rights. The attachment to blood lineage is a product of tribal rolls established by the

Bureau of Indian Affairs in the making of the reservation system in the US. This "attachment"—measured through the system of "blood quantum"—is a product of the history of settler colonialism with material consequences. Thanks to Ben Stork for providing us this history.

7. For an elaborate discussion of reparations and rights claims see Ta-Nehisi Coates (2014) and Vatter and De Leeuw (2023).
8. To define precision medicine is a tricky business. Some of the scholarship refers to any genetic technologies that come out of the human genome project, but others specify that precision goes beyond genetics and includes other "omics" such as proteomics or microbiomics. More recently, gene editing is also included in personalized forms of medicine, though methods here differ from the genome wide associations conducted in the "omics" space.
9. Singapore, for instance, recently established its national precision medicine initiative which focuses on advancing innovation in this area.
10. Compared to the group of patients who were taking the placebo, patients taking BiDil saw their mortality rate decrease by 43%. As a consequence, they stopped the trials so that the placebo group could take advantage of BiDil as well (for a further discussion see Dorr and Jones 2008).
11. Many thanks to Ben Stork for his suggestions in highlighting this dynamic.

References

Abel, S. 2018. What DNA Can't Tell: Problems with Using Genetic Tests to Determine the Nationality of Migrants. *Anthropology Today*, 34(6), pp. 3–6.

Anderson, B. 1991. *Imagined Communities: Reflections on the Origins and Spread of Nationalism.* London: Verso.

Arif, Y. 2016. Life, emergent: The social in the afterlives of violence. U of Minnesota Press.

Benjamin, Ruha. 2009. A Lab of Their Own: Genomic Sovereignty as Postcolonial Science Policy. *Policy and Society*, 28(4), pp. 341–355.

Bliss, C. 2012. *Race Decoded: The Genomic Fight for Social Justice.* Stanford: Stanford University Press.

Bolnick, D. A. Fullwiley, D. Duster, T. Cooper, R. S. Fujimura, J. H. Kahn, J. Kaufman, J. S. Marks, J. Morning, A. Nelson, A. and Ossorio, P. 2007. *The science and business of genetic ancestry testing. Science*, 318(5849), pp. 399–400.

Chalmers, D. Nicol, D. Otlowski, M. and Critchley, C. 2013. Personalised medicine in the genome era. J. Law Med, 20, pp. 577–594.

Chatterjee, P. 1993. *The Nation and Its Fragments: Colonial and Postcolonial Histories*, vol. 4. Princeton: Princeton University Press.

Chow-White, Peter A. 2012. The Informationalization of Race: Communication, Databases, and the Digital Coding of the Genome. In *Genetics and the Unsettled Past*, pp. 81–103. New Brunswick: Rutgers University Press.

Coates, Ta-Nehisi. 2014. The Case for Reparations. *Atlantic*. Available at https://www.theatlantic.com/magazine/archive/2014/06/the-case-for-reparations/361631/.

Collier, J. F., Maurer, B., and Suarez-Navaz, L. 1995. Sanctioned Identities: Legal Constructions of Modern Personhood. *Identities*, 2(1–2), pp. 1–27.

Davies, Margaret. 2023. Afterword. Feminist Legal Studies. https://doi.org/10.1007/s10691-022-09513-6.

Dickenson, D. 2013. Me medicine vs. we medicine: Reclaiming biotechnology for the common good. Columbia University Press.

Dickenson D, van Beers, B. and Sterckx, S. 2018. Introduction. In Van Beers B, van Beers BC, Sterckx S, Dickenson D, (eds) pp. 1–16, Personalised medicine, individual choice and the common good. Cambridge University Press.

Dorr, Gregory Michael, and Jones, David S. 2008. Facts and Fictions: BiDil and the Resurgence of Racial Medicine. *Journal of Law, Medicine and Ethics*, 36(3), pp. 443–448.

Dupras, C., Saulnier, K. M., and Joly, Y. 2019. Epigenetics, Ethics, Law and Society: A Multidisciplinary Review of Descriptive, Instrumental, Dialectical and Reflexive Analyses. *Social Studies of Science*, 49(5), pp. 785–810.

Dupré, J. 2008. What genes are and why there are no genes for race. *Revisiting race in a genomic age*, pp. 39–55.

Duster, T. 2006. *Backdoor to eugenics*. Routledge.

Duster, Troy. 2006. The Molecular Reinscription of Race: Unanticipated Issues in Biotechnology and Forensic Science. *Patterns of Prejudice*, 40(4–5), pp. 427–441.

Duster, T. 2015. A post-genomic surprise. The molecular reinscription of race in science, law and medicine. *The British journal of sociology*, 66(1), pp. 1–27.

El-Haj, N. A. 2012. *The genealogical science: The search for Jewish origins and the politics of epistemology*. University of Chicago Press.

Esposito, R. 2015. *Persons and things: From the bodies point of view*. John Wiley & Sons.

Esposito, Roberto. 2013. *Terms of the Political: Community, Immunity, Biopolitics: Community, Immunity, Biopolitics*. New York: Fordham University Press.

Fassin, Didier. 2013. The Precarious Truth of Asylum. *Public Culture*, 25(1), pp. 39–63.

Fassin, Didier and d'Halluin, Estelle. 2005. The Truth from the Body: Medical Certificates as Ultimate Evidence for Asylum Seekers. *American Anthropologist*, 107(4), pp. 597–608.

Faubion, James D., and Hamilton, Jennifer A. 2007. Sumptuary Kinship. *Anthropological Quarterly*, 80(2), pp. 533–559.

Fujimura, J. H. and Rajagopalan, R. 2011. Different differences: The use of 'genetic ancestry' versus race in biomedical human genetic research. *Social Studies of Science*, 41(1), pp. 5–30.

Fullwiley, Duana. 2007. The Molecularization of Race: Institutionalizing Human Difference in Pharmacogenetics Practice. *Science as Culture*, 16(1), pp. 1–30.

Gonzalez-Polledo, E. J. and Posocco, S. 2021. Archives, promises, values: forensic infrastructures in times of austerity. *Critique of Anthropology*, 41(1), pp. 3–20.

Green, S., Carusi, A., and Hoeyer, K. 2019. Plastic Diagnostics: The Remaking of Disease and Evidence in Personalized Medicine. *Social Science & Medicine*, 304, 112318.

Guthman, J. and Mansfield, B. 2013. The Implications of Environmental Epigenetics: A New Direction for Geographic Inquiry on Health, Space, and Nature-Society Relations. *Progress in Human Geography*, 37(4), pp. 486–504.

Habermas, J. 1996. *Between Facts and Norms*. Cambridge, MA: MIT Press.

Hacking, I. 2006. Genetics, Biosocial Groups & The Future of Identity. *Daedalus*, 135(4), pp. 81–95.

Hamilton, J. A. 2008. Revitalizing difference in the HapMap: Race and contemporary human genetic variation research. *Journal of Law, Medicine & Ethics*, 36(3), pp. 471–477.

Hamilton, Jennifer. 2012. The Case of the Genetic Ancestor. In Wailoo, K., Nelson, A., and Lee, C. (eds.), *Genetics and the Unsettled Past: The Collision of DNA, Race, and History*. New Brunswick: Rutgers University Press.

Hanafin, P. 2008. Voicing embodiment, relating difference: Towards a relational legal subjectivity. *Australian Feminist Law Journal*, 29(1), pp. 77–89.

Harris, C. 1993. Whiteness as Property. *Harvard Law Review*, 106(8), 1709.

Heinemann, T. and Lemke, T. 2014. Biological Citizenship Reconsidered: The Use of DNA Analysis by Immigration Authorities in Germany. *Science, Technology, & Human Values*, 39(4), pp. 488–510.

Hinchliffe, S. Bingham, N. Allen, J. and Carter, S. 2016. *Pathological lives: Disease, space and biopolitics*. John Wiley & Sons.

Hinterberger, A. 2012. Publics and populations: The politics of ancestry and exchange in genome science. *Science as Culture*, 21(4), pp. 528–549.

Hobsbawm, E. J. 1992. *Nations and Nationalism Since 1780: Programme, Myth, Reality*. Cambridge: Cambridge University Press.

Hoeyer, K. 2018. Lost and Found: Relocating the Individual in the Age of Intensified Data Sourcing in European Healthcare. *Personalized Medicine, Individual Choice and the Common Good*, pp. 144–166. Cambridge: Cambridge University Press.

Hoeyer, K. 2019. Data as Promise: Reconfiguring Danish Public Health Through Personalized Medicine. *Social Studies of Science*, 49(4), pp. 531–555.

Ingold, Tim and Palsson, Gisli, eds. 2013. *Biosocial Becomings: Integrating Social and Biological Anthropology*. Cambridge, UK: Cambridge University Press.

Ihar, Z. D. 2020. Phenotypic Personhood: Epigenetics and the Biolegality of Processing Asylum. In *Personhood in the Age of Biolegality*, pp. 127–147. Cham: Palgrave Macmillan.

Jasanoff, S., ed. 2004. *States of Knowledge: The Co-production of Science and the Social Order*. London: Routledge.

Jasanoff, S., ed. 2011. *Reframing Rights: Bioconstitutionalism in the Genetic Age*. Cambridge: MIT Press.

Jasanoff, S. 2016. Perfecting the Human: Posthuman Imaginaries and Technologies of Reason. In *Perfecting Human Futures*, pp. 73–95. Wiesbaden: Springer VS.

Joly, Yann, Braker, Maria, and Le Huynh, Michael. 2010. Genetic Discrimination in Private Insurance: Global Perspectives. *New Genetics and Society*, 29(4), pp. 351–368.

Jones, D. S. 2013. How Personalized Medicine Became Genetic, and Racial: Werner Kalow and the Formations of Pharmacogenetics. *Journal of the History of Medicine and Allied Sciences*, 68, pp. 1–48.

Kahn, J. 2004. How a Drug Becomes 'Ethnic': Law, Commerce, and the Production of Racial Categories in Medicine. *Yale Journal of Health Policy, Law, and Ethics*, 4(1), pp. 1–46.

Kahn, J. 2006. Genes, Race, and Population: Avoiding a Collision of Categories. *American Journal of Public Health*, 96(11), pp. 1965–1970.

Karpin, I. 2016. Regulatory responses to the gendering of transgenerational harm. *Australian Feminist Studies*, 31(88), pp. 139–153.

Kowal, E. and Watt, E. 2018. What is race in Australia. *Journal of Anthropological Sciences*, 96, pp. 229–237.

Landecker, H. and Panofsky, A. 2013. From Social Structure to Gene Regulation, and Back: A Critical Introduction to Environmental Epigenetics for Sociology. *Annual Review of Sociology*, 39, pp. 333–357.

Lappé, M. 2016. The maternal body as environment in autism science. *Social Studies of Science*, 46(5), pp. 675–700.

Lappé, M. Jeffries Hein, R. and Landecker, H. 2019. Environmental politics of reproduction. *Annual Review of Anthropology*, 48, pp. 133–150.

Lock, M. 2013. The Epigenome and Nature/Nurture Reunification: A Challenge for Anthropology. *Medical Anthropology*, 32 (4), pp. 291–308.

Lock, M. M. and Palsson, G. 2016. *Can Science Resolve the Nature/Nurture Debate?* New York: Wiley.

Manderson, L. 2016. Foetal politics and the prevention of chronic disease. *Australian Feminist Studies*, 31(88), pp. 154–171.

Mansfield, B. 2012. Race and the New Epigenetic Biopolitics of Environmental Health. *BioSocieties*, 7, pp. 352–372.

Mansfield, B. and Guthman, J. 2015. Epigenetic Life: Biological Plasticity, Abnormality, and New Configurations of Race and Reproduction. *Cultural Geographies*, 22(1), pp. 3–20.

M'charek, A. 2008. Silent Witness, Articulate Collective: DNA Evidence and the Inference of Visible Traits. *Bioethics*, 22(9), pp. 519–528.

M'charek, A. and Casartelli, S. 2019. Identifying Dead Migrants: Forensic Care Work and Relational Citizenship. *Citizenship Studies*, 23(7), pp. 738–757.

M'charek, A., Toom, V., and Jong, L. 2020. The Trouble with Race in Forensic Identification. *Science, Technology, & Human Values*, 45(5), pp. 804–828.

M'charek, A., Toom, V. and Prainsack, B. 2012. Bracketing off population does not advance ethical reflection on EVCs: A reply to Kayser and Schneider. *Forensic Science International: Genetics*, 6(1), pp. e16–e17.
Meloni, M. 2015. Epigenetics for the Social Sciences: Justice, Embodiment and Inheritance in the Postgenomic Age. *New Genetics and Society*, 34(2), pp. 125–151.
Meloni, M. 2017. Race in an Epigenetic Time: Thinking Biology in the Plural. *British Journal of Sociology*, 68(3), pp. 389–409.
Meloni, M. and Testa, G. 2014. Scrutinizing the Epigenetics Revolution. *BioSocieties*, 9, pp. 431–456.
Montoya, M. J. 2007. Bioethnic conscription: Genes, race, and Mexicana/o ethnicity in diabetes research. *Cultural anthropology*, 22(1), pp. 94–128.
Nash, C. 2005. Geographies of relatedness. *Transactions of the Institute of British Geographers*, 30(4), pp. 449–462.
Nelson, A. 2008. Bio science: Genetic genealogy testing and the pursuit of African ancestry. *Social studies of science*, 38(5), pp. 759–783.
Nelson, A. 2016. *The Social Life of DNA: Race, Reparations, and Reconciliation after the Genome*, Boston, MA: Beacon Press.
Nicol, D. Bubela, T. Chalmers, D. Charbonneau, J. Critchley, C. Dickinson, J. Fleming, J. Hewitt, A. W. Kaye, J. Liddicoat, J. and McWhirter, R., 2016. Precision medicine: drowning in a regulatory soup?. *Journal of Law and the Biosciences*, 3(2), pp. 281–303.
Niewöhner, J. 2011. Epigenetics: Embedded Bodies and the Molecularisation of Biography and Milieu. *BioSocieties*, 6(3), pp. 279–298.
Novas, C. and Rose, N. 2000. Genetic Risk and the Birth of the Somatic Individual. *Economy and Society*, 29(4), 485–513.
Nowotny, H. Scott, P. and Gibbons, M. 2001. *Re-thinking science: Knowledge and the public in an age of uncertainty* (p. 12). Cambridge: Polity.
Obasogie, O. K. 2008. Beyond best practices: strict scrutiny as a regulatory model for race-specific medicines. *Journal of Law, Medicine & Ethics*, 36(3), pp. 491–497.
O'Malley, M. A. and Skillings, D. J. 2018. Methodological Strategies in Microbiome Research and Their Explanatory Implications. *Perspectives on Science*, 26(2), 239–265.
Ong, A. 2010 'An Analytics of Biotechnology and Ethics at Multiple Scales', in Ong and Nancy N. Chen, eds, *Asian Biotech: Ethics and Communities of Fate* (Durham NC: Duke University Press, 2010).
Ong, A. 2016. *Fungible life: Experiment in the Asian city of life*. Duke University Press.

Ossorio, Pilar and Duster, Troy. 2005. Race and Genetics: Controversies in Biomedical, Behavioral, and Forensic Sciences. *American Psychologist*, 60(1), p. 115.

Ozdemir, V. Muljono, D. H. Pang, T. Ferguson, L. R. Manamperi, A., Samper, S. Someya, T. Tassé, A. M. Tsai, S. J. Zhou, H. H. and Lee, E. J. 2011. Asia-Pacific Health 2020 and genomics without borders: co-production of knowledge by science and society partnership for global personalized medicine. *Current Pharmacogenomics and Personalized Medicine*, 9(1), p. 1.

Pickersgill, M., Niewöhner, J., Müller, R., Martin, P. and Cunningham-Burley, S. 2013. Mapping the New Molecular Landscape: Social Dimensions of Epigenetics. *New Genetics and Society*, 32(4), pp. 429–447.

Pollock, Anne. 2008. Pharmaceutical Meaning-making Beyond Marketing: Racialized Subjects of Generic Thiazide. *Journal of Law, Medicine & Ethics*, 36(3), pp. 530–536.

Prainsack, B. 2017. Personalized medicine: empowered patients in the 21st century?. In Personalized Medicine. New York University Press.

Pugliese, J. 2012. *Biometrics: Bodies, Technologies, Biopolitics*. London: Routledge.

Rajagopalan, R. M., Nelson, A., and Fujimura, J. H. 2017. 12 Race and Science in the Twenty-First Century. In Felt, U., Fouché, R., Miller, C. A., and Smith-Doerr, L. (eds.), *The Handbook of Science and Technology Studies*, 4th ed., pp. 349–378. Cambridge: MIT Press.

Reardon, J. 2009. *Race to the Finish: Identity and Governance in an Age of Genomics*. Princeton University Press.

Reardon, Jenny. 2017. *The Postgenomic Condition: Ethics, Justice, and Knowledge After the Genome*. Chicago, IL: University of Chicago Press.

Reardon, Jenny and TallBear, Kim. 2012. "Your DNA Is Our History" Genomics, Anthropology, and the Construction of Whiteness as Property. *Current Anthropology*, 53(S5), S233–S245.

Roberts, D. E. 2006. Legal Constraints on the Use of Biomedical Research: Toward a Social Justice Framework. *Journal of Law, Medicine & Ethics*, 24(3), pp. 526–534.

Roberts, D. E. 2008. Is Race-based Medicine Good for Us? African American Approaches to Race, Biomedicine, and Equality. *Journal of Law, Medicine & Ethics*, 36(3), pp. 537–545.

Roberts, D. E. 2013. Law, race, and biotechnology: toward a biopolitical and transdisciplinary paradigm. *Annual Review of Law and Social Science*, 9, pp. 149–166.

Rose, N. 2007. Molecular Biopolitics, Somatic Ethics and the Spirit of Biocapital. *Social Theory & Health*, 5(1), pp. 3–29.

Rose, N. and Novas, C. 2005. Biological Citizenship. *Global Assemblages: Technology, Politics, and Ethics as Anthropological Problems*, pp. 439–463. Oxford: Blackwell Publishing.

Sankar P. 2010. Forensic DNA phenotyping: reinforcing race in law enforcement. *What's the Use of Race? Modern Governance and the Biology of Difference* I Whitmarsh, DS Jones 49–61 Cambridge, MA: MIT Press

Stern, Claudio D. 2019. The 'Omics Revolution: How an Obsession with Compiling Lists Is Threatening the Ancient Art of Experimental Design. *BioEssays*, 41(12), n/a–n/a. Web.

Sun, S. 2017. Socio-economics of personalized medicine in Asia, Routledge.

TallBear, K. 2013. Native American DNA: Tribal Belonging and the False Promise of Genetic Science. Minnesota: University of Minnesota Press.

Timmermans, S. and Buchbinder, M. 2010. Patients-in-waiting: living between sickness and health in the genomics era. *Journal of health and social behavior*, 51(4), pp. 408–423.

Valdez, N., 2018. The redistribution of reproductive responsibility: On the epigenetics of "environment" in prenatal interventions. *Medical Anthropology Quarterly*, 32(3), pp. 425–442.

Van Beers, B. C. 2018. The Best Me I Can Possibly Be: Legal Subjectivity, Self-Authorship and Wrongful Life Actions in an Age of 'Genomic Torts'. In *Personalised Medicine, Individual Choice and the Common Good*, pp. 200–225. Cambridge: Cambridge University Press.

Van Beers B, van Beers BC, Sterckx S, Dickenson D, (eds) 2018. *Personalised medicine, individual choice and the common good*. Cambridge University Press.

Van Wichelen, S. 2016. Postgenomics and Biolegitimacy: Legitimation Work in Transnational Surrogacy. *Australian Feminist Studies*, 31(88), pp. 172–186.

Van Wichelen, S. 2023. After Biosovereignty: The Material Transfer Agreement as Technology of Relations. *Social Studies of Science*, 53(4), pp. 599–621.

Van Wichelen, S. and Keaney, J. 2022. The Reproductive Bodies of Postgenomics. *Science, Technology, & Human Values*, 47(6), pp. 1111–1130.

Vatter, Miguel and De Leeuw, Marc. 2023. Human Rights, Legal Personhood and the Impersonality of Embodied Life. *Law, Culture and the Humanities*, 19(1), pp. 106–125.

Virruel-Fuentes, E. A. (2007). Beyond acculturation: Immigration, discrimination, and health research among Mexicans in the United States. *Social Science & Medicine*, 65, 1524–1535.

Vora, K. 2015. Re-imagining Reproduction: Unsettling Metaphors in the History of Imperial Science and Commercial Surrogacy in India. *Somatechnics*, 5(1), pp. 88–103.
Warin, M. Kowal, E. and Meloni, M. 2020. Indigenous knowledge in a postgenomic landscape: The politics of epigenetic hope and reparation in Australia. *Science, Technology, & Human Values*, 45(1), pp. 87–111.
Whitmarsh, Ian. 2009. Hyperdiagnostics: Postcolonial Utopics of Race-Based Biomedicine. *Medical Anthropology*, 28(3), pp. 285–315.
Whitmarsh, Ian and Jones, David S., eds. 2010. *What's the Use of Race? Modern Governance and the Biology of Difference*. New York: MIT Press.
Winickoff, D. E. and Obasogie, O. 2008. Race-Specific Drugs: Regulatory Trends and Public Policy. *Trends in Pharmacological Sciences*, 29(6), pp. 277–279.
Young, Iris Marion. 1990. *Justice and the Politics of Difference*. Princeton: Princeton University Press.

Epilogue: The Biolegality of Viruses

> There is a huge gulf between the state that is able to say 'I protect you from life and death,' that is to say from infection by a virus whose trace is known only to scientists and whose effects can only be understood by collecting statistics, and the state that would dare to say 'I protect you from life and death, because I maintain the conditions of habitability of all the living people on whom you depend'
>
> Bruno Latour, "Is This a Dress Rehearsal?"

Law Amidst Nature's Revenge

COVID-19 could be thought of as the perfect global biolegality crisis. As nature was "getting back at society," the life sciences were racing to find the ultimate technological fix in the form of a vaccine. Law—national, international, and local—was made on the spot, sending travelers, citizens, aliens, to the confines of their homes, neighborhoods, cities, and (nation) states. For the "greater good" human rights were forfeited. The aim was to destroy the virus and to secure the survival of as many people as possible. The race for vaccines not only freed up billions for science

S. van Wichelen and M. de Leeuw, *Biolegality*, Biolegalities,
https://doi.org/10.1007/978-981-99-8749-8

funding, but it also freed up laws and regulations to enforce medical testing, contact tracing, quarantine measures, lockdowns, and border closures. Both the biosciences and law took "a leap of faith" while populations questioned whether they should trust the science directing how states and governments would implement certain measures. On all levels, the COVID-19 pandemic revealed that bioscience today is deeply dependent on, embedded in, and determined by malleable legal regimes. Those legal regimes that allow, prohibit, or stimulate bioscientific actions create the "biolegal COVID-19 citizen"—as test-subject, contagious patient, subject under quarantine, or vaccine refuser. Like never before, the global pandemic engendered immunitary dispositifs; the synthesizing of nature sanctioned by masks, quarantines, lockdowns, vaccines, etc.

The biolegal framework of COVID-19 was visible on the macro- and micro-level of the pandemic. On the macro level, national governments and local states put their trust in the technological imaginary of bioscientists to solve the health crisis. Here, science and the state collaborated to protect the wellbeing of populations. On the micro-level, we all became biolegal subjects, testing ourselves in our domestic mini lab while distancing our bodies from others. As new "pop-up legal regimes" locked some people down in homes, and closed buildings and entire neighborhoods, others—such as essential workers—were forced to move and interact. Likewise, masks, pods, and spatial practices endorsed and enforced the enclosure of some families together while other (non-legal) kin arrangements suffered displacement and separation. Legal and biological knowledge justified this dual regime of arresting movement by cordoning off bodies to contain the virus. As states focused on how to keep the economy going amidst the pandemic, political scientists quickly pointed to the Foucauldian biopolitical governance of life or the production of Agamben's "bare life" as demonstrated by the political crisis ushered in by the pandemic. At the same time, epidemiologists mostly criticized failed public health systems made up of underfunded hospitals and underpaid nurses. Millions of people died, often alone, isolated from family and friends. The biolegal regime impacted our most intimate and painful moments as humans and communities (see also Lemm and Vatter 2022).

Yet as Bruno Latour reminds us in the opening quote of this chapter, this kind of biolegal regime "that seems to have come straight out of a Michel Foucault lecture" (p. S26 in 2021) does not precipitate a similar scenario for the climate crisis. Despite a stronger justification, there are no shutdowns of gas and oil companies, no mandates to ban pesticides, no intention—with the "supreme audacity" that closed beaches, playgrounds, and concert halls—to "ban outdoor heaters on bar terraces" (ibid). This is because, Latour exclaims, "the pathogen whose terrible virulence has changed the living conditions of all the inhabitants of the planet is not the virus at all—it is humanity!" As a "dress rehearsal" for the ecological crisis, COVID-19 failed to deliver: "If in the health crisis, it is the brave people who must relearn to wash their hands and cough into their elbows as they did in primary school, in the case of the ecological mutation, it is the state that finds itself in a learning situation" (p. S27 in Latour 2021). Whether preemptive or mitigating, the ecological challenges facing the planet today require not only new ethico-legal frameworks to prompt adequate measures and responses, but they also require new ways of "seeing like a state" (Scott 1998).

This epilogue identifies the biolegal regime constituted during the COVID-19 pandemic in order to understand how we might actively shape biolegal responses to future crises (including the climate crisis). More specifically, we look at the role of technology in prompting state-like decisions that implicate far reaching legal measures. Mathemetic disease modeling is such a technology, used ubiquitously during the latest pandemic. While population surveys (census) go back centuries as preferred tool to categorize, control, and govern citizens, and the use of mathematical models of infectious disease has been around since the late nineteenth century, current forms are much more complex and rely heavily on high computational power.[1] During the pandemic complex mathematical models were used to assess disease progression and the effects of different interventions (lock downs, mask requirements, vaccine rollouts), and therefore acquired unprecedented authority legitimized by scientific numbers enabling ad-hoc state policies and legal measures to be subjected to the public. Gathering more and more authority, modeling becomes integral to governance and demands a thorough biolegal analysis.

Modeling the Pandemic

In mid-March 2020, the head of the modeling team at Imperial College London, Neil Ferguson, presented the British state with scenarios where no serious intervention would lead to more than 510,000, mitigation measures to 250,000, and a suppression strategy to 20,000 COVID-19 deaths. At the time, the UK was practicing minor restrictions—believing that achieving herd-immunity would effectively bring cases down—and the information presented by the modeling team radically altered the politicians' response, triggering stringent lockdown measures.[2] The subsequent social and institutional call to "flatten the curve" illustrates the dominance mathematic disease modeling achieved during the pandemic. Despite models being contingent on context and the interplay of forces—and therefore not univocal, unequivocal, or transparent—their legitimacy held great sway in the political and public domain. Jones and Helmreich (2020) called the collective eagerness to flatten the curve a form of socio-technocratic optimism. Others similarly observed the way this ethos converted "science into social practices, calculations into materializations, abstracts into affects, and models into society" (see p. 255 in Rhodes et al., 2020). In all of this, modeling signaled the power of technology to change behavior. The COVID-19 pandemic presented the first opportunity to couple disease models with data beyond the public domain, including big data from a variety of platforms such as Google and Facebook.

According to sociologists Isin and Ruppert (2020), apps, devices, and platforms generate "the birth of sensory power," a new form of power distinct from—though also working within—the disciplinary and regulatory forms so assiduously described in Foucault's work on governmentality. Sensory power heavily depends on computational technologies and platform designs of sensors including those that have capacities to detect and to identify. Through the digitized data of sensors, the behavior and movements of individuals and groups are trackable and locatable, making them amenable for analysis, nudges, or control. COVID-19 epitomized this form of sensory power where sensory assemblages locate geographical hotspots and epicenters that Isin and Ruppert refer to as clusters: "If solutions were found to automate the testing and isolating stages, essentially multiple, relational, fluid, visualized and live clusters could be *governing themselves*" (see p. 11 in 2020, emphasis in original).

While Isin and Ruppert are critical of this development, the philosopher of technology Benjamin Bratton suggests that this new arrangement of power necessitates a form of "platform sovereignty," a mode of Foucauldian style governmentality that like other sites of governance (states, prisons, hospitals, schools, textbooks, discourse, technique, architecture) produces its own subjects and objects (Bratton 2016). As Bratton argues, platform sovereignty demands that we, "understand the designability of geography in relation to the designability of computation and to see the state (and other sovereign institutions) in relation to both at once" (see p. 27 in 2016). As a form of platform sovereignty, mathematic disease modeling governs and is governed by the specific platform technologies that organize and mobilize mathematical, digital, and algorithmic techniques. This requires a reliance not only on scientific quantitative reasoning, but, more significantly, a "scale of technology that comes to absorb functions of the state and the work of governance" (ibid). Thomas Levine—researching "modeling sovereignty" in China—describes how the computational infrastructures that facilitated disease modeling in China did indeed absorb state functions during the COVID-19 pandemic. He describes the way models "represent dynamic and live data to make analysis easier and decision-making faster" (see p. 70 in Levine 2022). As this creates new regulatory issues for the state and the law it is pertinent for us to consider such issues with biolegality in mind, pointing not only to capacities of modeling sovereignty, but also to their pitfalls.

The Sovereignty of Models

Both "platform sovereignty" and "modeling sovereignty" are unpalatable to those critical social science scholars trained to see technology in terms of surveillance, disruption of social cohesion, or as a form of negative biopolitics constraining social life. Sociologists in this vein rightly argued that the pandemic lockdowns (1) heightened surveillance of marginalized communities who often inhabited COVID-19 "hotspots," (2) significantly disrupted life for people working in certain industries, especially low wage sectors employing already impoverished and precarious populations, and (3) reproduced racialized and gendered vulnerabilities by framing biopolitical subjects around those who could stay at home and socially distance and those who could not and were left to die.

However, the global pandemic has propelled other social scientists—those moving away from critical analysis and toward multi- and cross-disciplinary research—to think *with* epidemiologists and to think *with* numbers. While they acknowledge that the deployment of numbers, data, and models are mediated and shaped by capital and global power relations, they follow the line of reasoning that the politics of climate change—just like the politics of pandemics—is more than the politics of capitalism (Chakrabarty 2007). According to this line of thought, viruses and climate change, ultimately, do not discriminate against humans and non-humans who are or will be affected (we humans do that).

There is enough evidence from the pandemic to counter this line of thought. The biolegality of COVID-19 became the "human condition"; the existential experience of life in pandemic times. Modeling like the kind Ferguson offered also produced a minimally acceptable level of death that assisted states in rationalizing their limited direct support for citizens (and for capitalist accumulation to remain unchallenged despite the lockdowns). While models endorsed vaccination as the primary technological fix, pharmaceutical companies gained unprecedented regulatory freedom and public financing to produce and put on the global market patented commodities in economically beneficial arrangements with states and governments. Northern states did not suspend international patent law to enable mass production of the mRNA vaccines globally that would have decreased the number of deaths worldwide. Hence, from North–South access and distribution of vaccines to ingrained fears of racialized minorities exacerbating the pandemic, the same biolegalities put in place to protect populations also put certain populations at risk.

An affirmative approach to biolegality does not deny or efface the biopolitical techniques that render and sustain the inequalities we see today and that hold up strong in times of crisis. On the contrary, a biolegal approach reveals these techniques as they are entangled in legal practice and knowledge. However, an affirmative approach attempts to focus actively on how to do things otherwise. Our critical introduction aspires to move beyond a total rejection of biopolitical techniques (à la Agamben) and to see how biolegal techniques might in fact, also contribute to the opposite of a necropolitics, offering instead, ways to use biolegal technologies to counter the unjust and unequal distribution of life and death, and of needs and responsibilities.[3]

However, the kind of platform sovereignty proposed does not—we contend—qualify as a viable affirmative form of biolegality. Three things prevent it from developing further into a legitimate form of governance (1) the struggle over social authority, (2) the difficulty of knowledge integration in systems thinking, and (3) the problem of scaling-up community participation.

The first complication, social authority, requires examining how authority is produced. This demands a close analysis of how and under what conditions states embrace the authority of mathematic modeling and which institutions are considered highly trustworthy by states and their experts to produce the numbers. When and how did modeling come to replace, rather than threaten, state authority? Warwick Anderson (1992) has looked at the struggle over authority in medical settings, describing how clinical scientists and computer experts represent their respective "skills" and "science" in a struggle over whether computing should be given diagnostic authority. Can we compare this situation with struggle over social authority between politicians and modelers early in the pandemic? If we were to understand modelers' assessments as a form of "diagnosis," then yes, the struggle for authority implies there is a certain context that affords different forms of authority. And in the early days of the pandemic the state accepted computational diagnosis, in the form of mathematical modeling, as socially legitimate. But as we know from how the pandemic progressed over time, this social legitimacy can wane. Toward the end of the pandemic, modeling was not enough for the state to grant the same kind of authority.[4] One can conclude here that the authority of modeling, in this instance, was countered and weakened. The struggles between modelers and other expert committees who rely on other methods or frameworks of evidence therefore represent a site where contesting competences are constituted.

In contrast to Anderson's story about the "craft" and "science" of diagnosis, this is not a story of clinical practice, but of state praxis. The difference lies in the fact that different "treatments" for managing the pandemic were integral to the "diagnostic" calculations, thus collapsing the order of a conventional clinical diagnosis. Modeled treatments become diagnosis of the social situation—for instance, acceptable levels of death and sickness. As such, strict lockdowns, mitigation measures, and laissez-faire approaches were among the "treatments" integrated into the calculations of mathematic modeling at the onset of the pandemic, just as mass vaccinations, strategic vaccinations, or no vaccinations were in the later debate around child vaccinations. The immanence of treatment to diagnosis in modeling made it difficult for different experts to compete for social authority; modeling incorporated and absorbed different treatments into their modeling programs, making them seem indispensable to state practice.

Second, mathematic modeling during the COVID-19 global pandemic had the potential to constitute new forms of authority, intervening in and affecting practices of *social* diagnosis and treatment. The question, though, is how and where authority is inscribed in the modeling systems (whether in the design phase or the actual practice of modeling). Social scientists argue that the knowledge systems feeding modeling should include more sociological, historical, geographical (spatial), and ecological knowledge (Anderson 2021 and 2023, Anderson and Dunk 2022, Hinchcliffe 2020). As many scholars point out, not only in relation to the pandemic but especially in relation to climate change, the social sciences are vital to the "making" of models, and more generally the "making" of "platform sovereignty" insofar as it ultimately deals with "state models." But what would this look like concretely? Modeling is ultimately a form of systems thinking and, in the mid-twentieth century, the highly abstract field of ecology, under the considerable influence of cybernetics and systems theory, incorporated mathematization as a foundational methodology. Systems thinking includes the interactions between organisms and physical and chemical processes in the environment. The problem is that these systems, as theorized or modeled, do not easily accommodate social—historical, philosophical, anthropological—parameters and inputs that social scientists work with. What to make of tacit knowledge—the kind of experiential knowledge difficult to quantify or communicate—or of sensitive ethnographic knowledge—the knowledge of cultural meanings acquired through long-term fieldwork and participant observation—that does not lend itself to a systems-thinking approach?

The third problem with endorsing a positive biopolitics of platform sovereignty is the way publics are either created by or interpellated in mathematic modeling. Social information either provided by anthropologists, ethnographers, and sociologists, or by communities themselves, is generally regarded as complementary to mathematical modeling, providing the sensitive/sensing qualities of context, history, and mood. Studies looking at these sensing knowledge inputs, however, differ when we compare epidemic and environmental modeling. Previous research on epidemics in developing countries, such as Vietnam and Indonesia in Southeast Asia or Monrovia and Liberia in West Africa, prompted epidemic modelers (often located in the global North or in metropolitan centers in the South), in response to infectious diseases like Ebola, H5N1, and SARS, to call upon social scientists—especially anthropologists—to assist with modeling. But while ethnographic details are useful in local medical humanitarian emergencies, they are rarely scaled up to inform models (Abramowitz et al. 2015). Social science involvement at modeling centers, such as the US Centers for Disease Control and Prevention, is marginal. Instead, what happens in public health is a top-down, rather than bottom-up, approach. Communities do not inform models; the modeling informs and directs behavioral changes in communities. Attempted nudging of anti-lockdown and anti-vaccine groups, and their irrationality, which we witnessed throughout the COVID-19 pandemic, exemplify this approach.[5]

Engaging the Planetary in Law

As Jones and Helmreich (2020) contend, the shape of epidemics emerges through data visualization, then evolves into an object of mathematical modeling, which finally morph into tools for intervention and governance. The role of disease modeling is helpful in thinking through contemporary problems in the science–law nexus, starting with how technology intervenes in the relationship between science and the state, particularly how technologies shape the legitimation and enactment of authority. This is followed by a second problem pertaining to what kind of legal knowledge certain technologies can afford. Thirdly there is the question of how technology mobilizes publics and activates forms of (anti)politics that challenge and/or alter the science–law nexus. While these are longstanding questions, the increased computational capacity to quickly process vast amounts of information, combined with the accelerated development of programming and the formation of new "platform economies," add new considerations to any attempt to fully understand the effect of technology on society and society's constitutional foundations.

Bruno Latour's words, with which we opened this chapter, rally scholars to vigilantly bring insights from the pandemic to bear on the climate crisis. Written at the height of the COVID-19 pandemic, the late philosopher laments the relationship between science and the state that persists in framing the pandemic as an anthropocentric problem, as a war on the virus. Comparing the health crisis with the climate crisis, he identifies a common diagnosis:

> What allows the two crises to occur in succession is the sudden and painful realization that the classical definition of society—humans among them-selves—makes no sense. The state of society depends at every moment on the associations among many actors, most of whom do not have human forms … The pandemic is no more a 'natural' phenomenon than the famines of the past or the current climate crisis. Society has long since moved beyond the narrow confines of the social sphere.

In combating SARS-COV-2, the state succumbed to a model of what the medical historian Charles Rosenberg called linear "contamination," which directs state response according to conventional assumptions around viral outbreaks that depend on disease progression devoid of any history or context (Rosenberg 1993; Anderson 2021). In following a contamination model, the state's response to COVID-19 was on the one hand, *not* a "dress-rehearsal" for the next crisis, as Latour so poignantly framed it. Mathematic modeling of climate change is not prompting states to intervene or declare states of emergency. On the other hand, COVID-19 was exactly a dress rehearsal for the next crisis. In establishing thresholds of acceptable levels and distributions of death, it projected the contours of their next—or even current—response. Worldwide, the poor and marginalized were overrepresented among the dead, just as they are in climate catastrophes whether in terms of bushfires, floods, hurricanes, and heat waves. Inequality between the North and South, between the rich and poor, continues to be ignored in the global response to climate change.

Accepting the Anthropocene thesis in all its critical forms, Latour's proposal to observe the pandemic as a human—rather than natural—phenomenon, necessitates a different model, namely that of "configuration," which encompasses a more complex, ecological, and heterogenous understanding of epidemics drawing in knowledge from the biological *and* the social sciences (Rosenberg 1993, see also Anderson 2021). Hence, the lesson of the ongoing climate change crisis is that climate action must be based on a model of configuration that considers biosocial approaches to life and the polis. What law needs to be in this current predicament—moving into an all-encompassing crisis—is to be attuned to a model of configuration and a commitment to reconfigure law.

Such a commitment necessitates keeping law central to the analysis of biopolitics. Legal anthropologists are already at the forefront of demonstrating how law connects people and their environments. Our study of genes, brains, babies, bodies, and viruses, shows how individuals and communities appeal to legal institutions to pursue life affirming situations, with respect to both human and more-than-human relations. At the same time, we illustrated of how and where biolegalities often fall short of justice. This condition evokes the central legal paradox that "law is required in order for there to be [...] justice, yet law cannot deliver the justice that it promises" (see p. 88 in Greenhouse 2008). Here, Greenhouse reminds us not only of the multivalent relationship between law and justice (Coutin 2020), but also of the enigmatic dynamic between nature and biology. Going forward it is essential to keep in mind that "neither law's presence nor law's absence can bring about heaven, but the quest itself, 'holding open every possible question' is worthwhile" (see par 12 in Coutin 2020). We bring our book to an end on this note and with the knowledge that more work on biolegality lies ahead and that we have, hopefully, contributed to setting the scene and tone for this labor.

Notes

1. Conventional forms of mathematic disease modeling rely on compartmental models that chart the mathematical relationship between groups (compartments) of the population. One dominant compartmental model is the SIR model, which groups those who are susceptible to the disease (S), those who are infected (I), and those who are Recovered (R). The SIR model has been successful in modeling disease progression, the number of infections, and calculating the so-called "basic reproduction number," or "R-nought" (R0). R0 "represents the number of new infections estimated to stem from a single case" (Fischer 2020). When the R0 runs below 1 the model indicates that the number of cases is shrinking. An R0 above 1 indicates growth of cases and during the COVID-19 pandemic these instances called for public interventions, such as masking or lockdowns.
2. This was not without challenge. The sudden change of government response prompted much criticism. One major assessment was how modeling was given too much agency (Friedman 2020). While Ferguson's modeling was key to the lockdowns it was not without a struggle for legitimacy.
3. Many thanks to Ben Stork for adding these clarities around our formulations.
4. At one point, the UK, for instance, became cautious about advocating vaccines for children—both for children between 12 to 17 years old as well as for children between 5 and 12 years old. The UK government's independent vaccine advisers, the Joint Committee on Vaccination and Immunization (JCVI), recommended against COVID vaccines for healthy teenagers despite calculations from mathematic modeling coming out of Warwick and Public Health England (PHE) that "fully vaccinating 80% of 12- to 17-year-olds would lead to 'large reductions' in infections and a 'substantial reduction' in hospitalizations in the age group" (Sample 2021).
5. By contrast, so-called "community-modeling" in environmental models has scaled up, for instance in the Community Multi-Scale Air Quality (CMAQ) modeling system and has been instrumental in shaping a pedagogic capacity for communities to impact both modeling and recommendations (Bigras

2013). It has also helped to form "respiratory publics" in China (Nguyen 2020). This suggests a bottom-up approach where communities—when taught how to contribute to modeling—can in some way not only influence the course of modeling but also illustrate the legitimacy and epistemic value of modeling to other publics.

References

Abramowitz, S. A., Bardosh, K. L., Leach, M., Hewlett, B., Nichter, M., and Nguyen, V. K. 2015. Social Science Intelligence in the Global Ebola Response. *The Lancet*, 385(9965), p. 330.

Anderson, B. 1991. *Imagined Communities: Reflections on the Origins and Spread of Nationalism*. London: Verso.

Anderson, W. 2021. The Model Crisis, or How to Have Critical Promiscuity in the Time of Covid-19. *Social Studies of Science*, 51(2), pp. 167–188.

Anderson, W. 2023. Toward Planetary Health Ethics? Refiguring Bios in Bioethics. *Journal of Bioethical Inquiry*, pp. 1–8. https://doi.org/10.1007/s11673-023-10285-0.

Anderson, W. and Dunk, J. 2022. Planetary Health Histories: Toward New Ecologies of Epidemiology? *Isis*, 113(4), pp. 767–788.

Bratton, B. H. 2016. *The Stack: On Software and Sovereignty*. Cambridge: MIT Press.

Chakrabarty, D. 2017. The Politics of Climate Change Is More Than the Politics of Capitalism. *Theory, Culture & Society*, 34(2–3), pp. 25–37.

Coutin, Susan B. 2020. The Cultural Prerequisites of Law: Engaging with the Work of Carol Greenhouse Over Time. *PoLAR: Political and Legal Anthropology Review*, July 5. https://politicalandlegalanthroreview.wordpress.com/2020/07/05/the-cultural-prerequisites-of-law-engaging-with-the-work-of-carol-greenhouse-over-time/?preview_id=10828&preview_nonce=d504a03345&post_format=standard&_thumbnail_id=10884&preview=true.

Fischer, M. 2020. R0, the Messy Metric That May Soon Shape Our Lives, Explained, *The New York Times*, April 23. https://www.nytimes.com/2020/04/23/world/europe/coronavirus-R0-explainer.html.

Friedman, J. 2020. PM Finally Defers to Experts as He Deflects Tricky Covid Questions. *The Guardian*, March 17. https://www.theguardian.com/politics/2020/mar/16/johnson-finally-defers-to-experts-as-he-deflects-tricky-covid-questions.

Greenhouse, Carol J. 2008. Life Stories, Law's Stories: Subjectivity and Responsibility in the Politicization of the Discourse of 'Identity.' *PoLAR: Political and Legal Anthropology Review*, 31(1) (May), pp. 79–95.

Jasanoff, S. 2020. Pathologies of Liberty: Public Health Sovereignty and the Political Subject in the Covid-19 Crisis. *Cahiers Droit, Sciences & Technologies*, 11, pp. 125–149.

Jones, David S. and Helmreich, Stefan. 2020. The Shape of Epidemics. *Boston Review*, June 26. https://www.bostonreview.net/articles/david-shumway-jones-stefan-helmreich-epidemic-waves/.

Latour, B. 2021. Is This a Dress Rehearsal? *Critical Inquiry*, 47(S2), S25–S27.

Lemm, V. and Vatter, M. eds., 2022. *The Viral Politics of Covid-19: Nature, Home, and Planetary Health*. Springer Nature.

Levine, I. T. 2022. Modeling Sovereignty: Towards a Legal Philosophy for Platforms in the People's Republic of China. *Peking University Law Journal*, 10(1), pp. 25–73.

Nguyen, V. 2020. Breathless in Beijing: Aerial Attunements and China's New Respiratory Publics. *Engaging Science, Technology, and Society*, 6, pp. 439–461.

Rhodes, T., Lancaster, K., and Rosengarten, M. 2020. A Model Society: Maths, Models and Expertise in Viral Outbreaks. *Critical Public Health*, 30(3), pp. 253–256.

Rosenberg, C. E. 1993. *Explaining Epidemics*. Cambridge: Cambridge University Press.

Sample, I. 2021. JCVI Failed to Back Youth Covid Jabs Despite Favourable Modelling. *The Guardian*, November 3. https://www.theguardian.com/politics/2021/nov/02/jcvi-failed-back-youth-covid-jabs-despite-favourable-models.

Scott, J. C. 1998. *Seeing Like a State: How Certain Schemes to Improve the Human Condition Have Failed*. New Haven, CT: Yale University Press.

References

Abel, S. 2018. What DNA Can't Tell: Problems with Using Genetic Tests to Determine the Nationality of Migrants. *Anthropology Today*, 34(6), pp. 3–6.

Abramowitz, S. A., Bardosh, K. L., Leach, M., Hewlett, B., Nichter, M., and Nguyen, V. K. 2015. Social Science Intelligence in the Global Ebola Response. *The Lancet*, 385(9965), p. 330.

Agamben, G. 1995. *Homo sacer*, p. 46. Torino: Einaudi.

Alac, M. 2008. Working with Brain Scans: Digital Images and Gestural Interaction in FMRI Laboratory. *Social Studies of Science*, 38(4), pp. 483–508.

Almeling, R. 2011. *Sex Cells: The Medical Market for Eggs and Sperm.* Berkeley: University of California Press.

Amendola, A. 2012. The Law of the Living: Material for Hypothesizing the Biojuridical. *Law, Culture and the Humanities*, 8(1), pp. 102–118.

Anderson, B. 1991. *Imagined Communities: Reflections on the Origins and Spread of Nationalism.* London: Verso.

Anderson, Warwick. 2008. *The Collectors of Lost Souls: Turning Kuru Scientists into Whitemen.* Baltimore: Johns Hopkins University Press.

Anderson, W. 2000. The possession of kuru: medical science and biocolonial exchange. *Comparative Studies in Society and History*, 42(4), pp.713–744.

S. van Wichelen and M. de Leeuw, *Biolegality*, Biolegalities,
https://doi.org/10.1007/978-981-99-8749-8

Anderson, W. 2021. The Model Crisis, or How to Have Critical Promiscuity in the Time of Covid-19. *Social Studies of Science*, 51(2), pp. 167–188.

Anderson, W. 2023. Toward Planetary Health Ethics? Refiguring Bios in Bioethics. *Journal of Bioethical Inquiry*, pp. 1–8. https://doi.org/10.1007/s11673-023-10285-0.

Anderson, W. and Dunk, J. 2022. Planetary Health Histories: Toward New Ecologies of Epidemiology? *Isis*, 113(4), pp. 767–788.

Andorno, R., Baylis, F., Darnovsky, M., Dickenson, D., Haker, H., Hasson, K., … and Zuscinova, J. 2020. Geneva Statement on Heritable Human Genome Editing: The Need for Course Correction. *Trends in Biotechnology*, 38(4), pp. 351–354.

Ankeny, R. and Leonelli, S. 2015. Valuing data in postgenomic biology. In Richardson, S.S. and Stevens, H. eds. *Postgenomics: Perspectives on biology after the genome*. pp. 126–49, Durham: Duke University Press.

Asdal, K., Druglitro, T., and Hinchliffe, S., eds. 2016. *Humans, Animals and Biopolitics: The More-Than-Human Condition.* Routledge.

Ardill, Allan. 2009. Sociobiology, Racism and Australian Colonisation. *Griffith Law Review,* 18(1), pp. 82–113. https://doi.org/10.1080/10383441.2009.10854631.

Arif, Y. 2016. Life, emergent: The social in the afterlives of violence. U of Minnesota Press.

Baccarini, Elvio and Malatesti, Luca. 2017. The Moral Bioenhancement of Psychopaths. *Journal of Medical Ethics*, 43(10), pp. 697–701.

Barrangou, R. and Doudna, J. A. 2016. Applications of CRISPR Technologies in Research and Beyond. *Nature Biotechnology*, 34(9), pp. 933–941.

Beaulieu, A., & Leonelli, S. 2021. *Data and Society: A Critical Introduction.* SAGE.

Benjamin, Ruha. 2009. A Lab of Their Own: Genomic Sovereignty as Postcolonial Science Policy. *Policy and Society*, 28(4), pp. 341–355.

Benjamin, R. 2015. The Emperor's New Genes: Science, Public Policy, and the Allure of Objectivity. *The ANNALS of the American Academy of Political and Social Science*, 661(1), pp. 130–142.

Bernstein, S. and Wiesemann, C. 2014. Should Postponing Motherhood Via "Social Freezing" be Legally Banned? An Ethical Analysis. *Laws*, 3(2), pp. 282–300.

Besson, S. 2007. Enforcing the Child's Right to Know Her Origins: Contrasting Approaches Under the Convention on the Rights of the Child and the European Convention on Human Rights. *International Journal of Law, Policy and the Family*, 21, pp. 137–159.

Biagioli, M. 2006. Patent Republic: Representing Inventions, Constructing Rights and Authors. *Social Research: An International Quarterly*, 73(4), pp. 1129–1172.

Biagioli, M. 2011. Patent Specification and Political Representation: How Patents Became Rights. *Making and Unmaking Intellectual Property: Creative Production in Legal and Cultural Perspective*, pp. 25–40. Chicago: University of Chicago Press.

Biagioli, M. and Buning, M. 2019. Technologies of the Law/ Law as a Technology. *History of Science*, 57(1), pp. 3–17.

Biagioli, M. and Pottage, A. 2021. Patenting Personalized Medicine: Molecules, Information, and the Body. *Osiris*, 36(1), pp. 221–240.

Bigenwald, A. and Chambon, V. 2019. Criminal Responsibility and Neuroscience: No Revolution Yet. *Frontiers in Psychology*, 10, p. 1406.

Bigras, E. 2019. Can the Model Teach? *The Asthma Files*. Platform for Experimental Collaborative Ethnography, Contributed by Mike Fortun, last modified 17 July 2019. Accessed 7 November 2021. Available at https://theasthmafiles.org/content/can-model-teach.

Bigras, E. 2020. *Can the Model Teach? Environmental Governance Through Technical Practice, theasthmafiles.org*. https://theasthmafiles.org/sites/default/files/artifacts/media/pdf/can_the_model_teach_environmental_govern.pdf.

Bliss, C. 2012. *Race Decoded: The Genomic Fight for Social Justice.* Stanford: Stanford University Press.

Bolnick, D. A. Fullwiley, D. Duster, T. Cooper, R. S. Fujimura, J. H. Kahn, J. Kaufman, J. S. Marks, J. Morning, A. Nelson, A. and Ossorio, P. 2007. *The science and business of genetic ancestry testing. Science*, 318(5849), pp.399–400.

Bourdieu, P. 1986. The Force of Law: Toward a Sociology of the Juridical Field. *Hastings Law Journal*, 38, p. 805.

Bowcott, O. 2016. Couple win legal battle against ruling on dead daughter's eggs, The Guardian, 30 June 2016, https://www.theguardian.com/uk-news/2016/jun/30/couple-win-leave-to-appeal-against-ruling-on-dead-daughters eggs

Bowker, G. C. and Star, S. L. 2000. *Sorting Things Out: Classification and Its Consequences*. Cambridge: MIT Press.

Braithwaite, J. and Drahos, P. 2000. *Global Business Regulation.* Cambridge university press.

Bratton, B. 2021. *The Revenge of the Real: Politics for a Post-pandemic World*. London: Verso Books.

Bratton, B. H. 2016. *The Stack: On Software and Sovereignty*. Cambridge: MIT Press.

Braverman, I., ed. 2017. *Gene Editing, Law, and the Environment: Life Beyond the Human*. London: Routledge.

Brown, W. 2002. *Moralism as Antipolitics Materializing Democracy: Toward a Revitalized Cultural Politics*, pp. 368–392. London: Duke University Press.

Brown, Wendy. 2019. *In the Ruins of Neoliberalism: The Rise of Antidemocratic Politics in the West*. New York: Columbia University Press.

Budiastuti, A. 2017. In DNA We Trust? Biolegal Governmentality and Illegal Logging in Contemporary Indonesia. *East Asian Science, Technology and Society: An International Journal*, 11(1), pp. 51–70.

Butler, Judith. 2002. *Gender Trouble*. London and New York: Routledge.

Butler, T., Schofield, P. W., Knight, L., Ton, B., Greenberg, D., Scott, R. J., Grant, L., Keech, A. C., Gebski, V., Jones, J., Ellis, A., Weatherburn, D., Wilhelm, K., Jones, A., Churchill, A., Allnutt, S., Mitchell, P. B., Chappell, D., Deste, C., Villa, D., and Carr, V. 2021. Sertraline Hydrochloride for Reducing Impulsive Behaviour in Male, Repeat-Violent Offenders (ReINVEST): Protocol for a Phase IV, Double-blind, Placebo-Controlled, Randomised Clinical Trial. *BMJ Open*, 11, pp. e044656. https://doi.org/10.1136/bmjopen-2020-044656.

Bühl, Lukas, Malarciuc, Christian, and Völlmecke, Anna. 2015. Transatlantic Differences in GMO Regulation. MaRBLe 4.

Cahn, Naomi R. 2009. *Test Tube Families: Why the Fertility Market Needs Legal Regulation*. New York: NYU Press.

Callaway, E. 2013. Deal Done Over Hela Cell Line. *Nature*, 500(7461), pp. 132–133.

Callon, M. 1994. Is Science a Public Good? Fifth Mullins Lecture, Virginia Polytechnic Institute, March 23, 1993. *Science, Technology, & Human Values*, 19(4), pp. 395–424.

Callon, Michel. 1998. Introduction: The Embeddedness of Economic Markets in Economics. *The Sociological Review*, 46(1), pp. 1–57.

Campbell, T. and Luisetti, F. 2010. On Contemporary French and Italian Political Philosophy: An Interview with Roberto Esposito. *Minnesota Review*, 75(1), pp. 109–118.

Caplan, Lincoln. 2020. Should Billy Joe Wardlow be Executed for a Crime Committed When He Was Eighteen? *The New Yorker*, June 30. https://www.newyorker.com/news/news-desk/should-billy-joe-wardlow-be-executed-for-a-crime-committed-when-he-was-eighteen.

Carolan, M. S. 2010. The Mutability of Biotechnology Patents: From Unwieldy Products of Nature to Independent 'Object/s'. *Theory, culture & society*, 27(1), pp. 110–129.

Carroll, K. and Kroløkke, C. 2018. Freezing for Love: Enacting 'Responsible' Reproductive Citizenship Through Egg Freezing. *Culture, Health & Sexuality*, 20(9), pp. 992–1005.

Carsten, Janet, ed. 2000. *Cultures of Relatedness: New Approaches to the Study of Kinship*. Cambridge: Cambridge University Press.

Chakrabarty, D. 2017. The Politics of Climate Change Is More Than the Politics of Capitalism. *Theory, Culture & Society*, 34(2–3), pp. 25–37.

Chalmers, D. Nicol, D. Otlowski, M. and Critchley, C. 2013. Personalised medicine in the genome era. J. Law Med, 20, pp.577–594.

Chang, Z., Lichtenstein, P., Langstrom, N., et al. 2016. Association Between Prescription of Major Psychotropic Medications and Violent Reoffending After Prison Release. *JAMA*, 316(17), pp. 1798–1807.

Chatterjee, P. 1993. *The Nation and Its Fragments: Colonial and Postcolonial Histories*, vol. 4. Princeton: Princeton University Press.

Chow-White, Peter A. 2012. The Informationalization of Race: Communication, Databases, and the Digital Coding of the Genome. In *Genetics and the Unsettled Past*, pp. 81–103. New Brunswick: Rutgers University Press.

Cloatre, E. 2018. Law and ANT (and Its Kin): Possibilities, Challenges, and Ways Forward. *Journal of Law and Society*, 45(4), pp. 646–663.

Coates, Ta-Nehisi. 2014. The Case for Reparations. *Atlantic*. Available at https://www.theatlantic.com/magazine/archive/2014/06/the-case-for-reparations/361631/.

Cohen, M. 2019. The Law of Placenta. *Yale Journal of Law and Feminism*, 31, p. 337.

Collier, Stephen J. and Lakoff, Andrew. 2015. Vital Systems Security: Reflexive Biopolitics and the Government of Emergency. *Theory, Culture & Society*, 32(2), pp. 19–51.

Collier, J. F., Maurer, B., and Suarez-Navaz, L. 1995. Sanctioned Identities: Legal Constructions of Modern Personhood. *Identities*, 2(1–2), pp. 1–27.

Cooper, M. E., 2008. *Life as surplus: Biotechnology and capitalism in the neoliberal era*. University of Washington Press.

Cooper, M. E. 2011. *Life as Surplus: Biotechnology and Capitalism in the Neoliberal Era*. Washington: University of Washington Press.

Cooper, M. and Waldby, C. 2014. *Clinical labor: Tissue donors and research subjects in the global bioeconomy*. Duke University Press.

Cotterrell, R. 1992. Law's Community: Legal Theory and the Image of Legality. *Journal of Law and Society*, 19(4), 405–422.
Cotterrell, R.B. 1992. *The Politics of Jurisprudence: A Critical Introduction to Legal Philosophy*. University of Pennsylvania Press.
Coutin, Susan B. 2020. The Cultural Prerequisites of Law: Engaging with the Work of Carol Greenhouse Over Time. *PoLAR: Political and Legal Anthropology Review*, July 5. https://politicalandlegalanthroreview.wordpress.com/2020/07/05/the-cultural-prerequisites-of-law-engaging-with-the-work-of-carol-greenhouse-over-time/?preview_id=10828&preview_nonce=d504a03345&post_format=standard&_thumbnail_id=10884&preview=true.
Cripps, Y. 2018. The Global Person: Pig-Human Embryos, Personhood, and Precision Medicine. *Indiana Journal of Global Legal Studies*, 25(2), pp. 701–727.
D'Alton-Harrison, R. 2014. Mater Semper Incertus Est: Who's Your Mummy? *Medical Law Review*, 22(3), pp. 357–383.
Davies, K. and White, M. 1996. *Breakthrough: The Race to Find the Breast Cancer Gene*. New York: Wiley.
Davies, M. 2020. Distributed Cognition Distributed Being, and the Foundations of Law. In *Personhood in the Age of Biolegality*, pp. 205–223. Cham: Palgrave Macmillan.
Davies, Margaret. 2023. Afterword. Feminist Legal Studies. https://doi.org/10.1007/s10691-022-09513-6.
Delaney, D. 2003. *Law and Nature*. Cambridge: Cambridge University Press.
De Leeuw, M. and Van Wichelen, S. 2020. *Brave New Law: Personhood in the Age of Biolegality*. Cham: Palgrave Macmillan.
De Waal, Frans. 2005. *Our Inner Ape*. New York: Riverhead Books.
Dickenson, D. 2013. Me medicine vs. we medicine: Reclaiming biotechnology for the common good. Columbia University Press.
Dickenson D, van Beers, B. and Sterckx, S. 2018. Introduction. In Van Beers B, van Beers BC, Sterckx S, Dickenson D, (eds) pp. 1–16, Personalised medicine, individual choice and the common good. Cambridge University Press.
Diver, A. 2014. *A Law of Blood-Ties-the 'Right' to Access Genetic Ancestry*. London: Springer International Publishing.
Dolgin, J. L. 1997. *Defining the Family: Law, Technology, and Reproduction in an Uneasy Age*. New York: NYU Press.

Dorr, Gregory Michael, and Jones, David S. 2008. Facts and Fictions: BiDil and the Resurgence of Racial Medicine. *Journal of Law, Medicine and Ethics*, 36(3), pp. 443–448.

Doudna, J. A. and Steinberg, S. H. 2017. A Crack in Creation: Gene Editing and the Unthinkable Power to Control Evolution. *The FASEB Journal*, 32(1), pp. 3–4.

Drahos, Peter and Braithwaite, John 2001. Intellectual Property, Corporate Strategy, Globalisation: TRIPS in Context. *Wisconsin International Law Journal*, 20, p. 451.

Dumit, Joseph. 2004. *Picturing Personhood: Brain Scans and Biomedical Identity*. Princeton: Princeton University Press.

Dumit, Joseph. 2012. Critically Producing Brain Images of Mind. In Choudhury, S. and Slaby, J. (eds.), *Critical Neuroscience: A Handbook of the Social and Cultural Contexts of Neuroscience*, pp. 195–225. Malden: Wiley-Blackwell.

Dupré, J. 2008. What genes are and why there are no genes for race. *Revisiting race in a genomic age*, pp.39–55.

Dupras, C., Saulnier, K. M., and Joly, Y. 2019. Epigenetics, Ethics, Law and Society: A Multidisciplinary Review of Descriptive, Instrumental, Dialectical and Reflexive Analyses. *Social Studies of Science*, 49(5), pp. 785–810.

Duster, T. 2006. *Backdoor to eugenics*. Routledge.

Duster, Troy. 2006. The Molecular Reinscription of Race: Unanticipated Issues in Biotechnology and Forensic Science. *Patterns of Prejudice*, 40(4–5), pp. 427–441.

Duster, T. 2015. A post-genomic surprise. The molecular reinscription of race in science, law and medicine. *The British journal of sociology*, 66(1), pp. 1–27.

Dworkin, R. M. 1993. *Life's Dominion: An Argument About Abortion, Euthanasia, and Individual Freedom*. New York: Vintage.

El-Haj, N. A. 2012. *The genealogical science: The search for Jewish origins and the politics of epistemology*. University of Chicago Press.

Elliott, E. D. 1984. Holmes and Evolution: Legal Process as Artificial Intelligence. *The Journal of Legal Studies*, 13(1), pp. 113–146.

Elliott, D.E. 1997. Law and biology: the new synthesis. *Saint Louis University Law Journal*, *41*(2), pp. 595–624.

Esposito, Roberto. 2013. *Terms of the Political: Community, Immunity, Biopolitics: Community, Immunity, Biopolitics*. New York: Fordham University Press.

Esposito, R. 2015. *Persons and Things: From the Body's Point of View.* New York: Wiley.

Esvelt, K. M. 2017. Rules for Sculpting Ecosystems: Gene Drives and Responsive Science. In *Gene Editing, Law, and the Environment*, pp. 20–37. London: Routledge.

Everett, M. 2007. The 'i' in the Gene: Divided Property, Fragmented Personhood, and the Making of a Genetic Privacy Law. *American Ethnologist*, 34(2), pp. 375–386.

Ewick, P. and Silbey, S. S. 1998. *The Common Place of Law: Stories from Everyday Life*. Chicago: University of Chicago Press.

Fabian, J. 2014. *Time and the Other*. New York: Columbia University Press.

Farahany, N. A. 2015. Neuroscience and Behavioral Genetics in US Criminal Law: An Empirical Analysis. *Journal of Law and the Biosciences*, 2(3), pp. 485–509.

Fassin, D. 2009. Another Politics of Life Is Possible. *Theory, Culture & Society*, 26(5), pp. 44–60.

Fassin, Didier. 2013. The Precarious Truth of Asylum. *Public Culture*, 25(1), pp. 39–63.

Fassin, Didier and d'Halluin, Estelle. 2005. The Truth from the Body: Medical Certificates as Ultimate Evidence for Asylum Seekers. *American Anthropologist*, 107(4), pp. 597–608.

Faubion, James D., and Hamilton, Jennifer A. 2007. Sumptuary Kinship. *Anthropological Quarterly*, 80(2), pp. 533–559.

Fearnley, Lyle. 2020. Viral Sovereignty or Sequence Etiquette? Asian Science, Open Data, and Knowledge Control in Global Virus Surveillance. *East Asian Science, Technology and Society: An International Journal*, 14(3), pp. 479–505.

Felt, U., Wynne, B., Callon, M., Gonçalves, M. E., Jasanoff, S., Jepsen, M., Joly, P. B., Konopasek, Z., May, S., and Neubauer, C. 2007. *Taking European Knowledge Society Seriously*. p. 700. Luxembourg: DG for Research.

Finkler, K. 2000. *Experiencing the New Genetics: Family and Kinship on the Medical Frontier*. Philadelphia: University of Pennsylvania Press.

Finnis, John. 2020. Natural Law Theories. In Zalta, E. N. (ed.), *The Stanford Encyclopedia of Philosophy* (Summer 2020 ed.). https://plato.stanford.edu/archives/sum2020/entries/natural-law-theories/.

Fischer, M. 2020. R0, the Messy Metric That May Soon Shape Our Lives, Explained, *The New York Times*, April 23. https://www.nytimes.com/2020/04/23/world/europe/coronavirus-R0-explainer.html.

Foster, L. A., 2017. *Reinventing Hoodia: Peoples, Plants, and Patents in South Africa*. University of Washington Press.

Foucault, M. 1976. *The Will to Knowledge: The History of Sexuality*, vol. 1 (trans. R. Hurley, 1998). London: Penguin.

Foucault, M. 1977. *Discipline and Punish: The Birth of the Prison* (trans. A Sheridan). London: Allen Lane.

Foucault, M. 2000. Truth and Power. In Faubion, J. D. (ed.), *Michel Foucault: Power*, pp. 111–133. New York: The New Press.

Franklin, S. 1995. Science as Culture, Cultures of Science. *Annual Review of Anthropology*, 24(1), pp. 163–184.

Franklin, S. 2001. Biologization Revisited: Kinship Theory in the Context of the New Biologies. In Franklin, S. and McKinnon, S. (eds.), *Relative Values: Reconfiguring Kinship Studies*, pp. 302–325. Durham: Duke University Press.

Franklin, S. 2006. The Cyborg Embryo: Our Path to Transbiology. *Theory, Culture, and Society*, 23(7–8), pp. 167–187.

Franklin, S. 2013. *Biological Relatives: IVF, Stem Cells and the Future of Kinship*. Durham: Duke University Press.

Friedman, J. 2020. PM Finally Defers to Experts as He Deflects Tricky Covid Questions. *The Guardian*, March 17. https://www.theguardian.com/politics/2020/mar/16/johnson-finally-defers-to-experts-as-he-deflects-tricky-covid-questions.

Fujimura, J. H. and Rajagopalan, R. 2011. Different differences: The use of 'genetic ancestry'versus race in biomedical human genetic research. *Social Studies of Science*, 41(1), pp. 5–30.

Fukuyama, F. 2003. *Our Posthuman Future: Consequences of the Biotechnology Revolution*. New York: Farrar, Straus and Giroux.

Fullwiley, Duana. 2007. The Molecularization of Race: Institutionalizing Human Difference in Pharmacogenetics Practice. *Science as Culture*, 16(1), pp. 1–30.

Gallagher, J. 2015. MPs Say Yes to Three-Person Babies. *BBC News*, February 3. https://www.bbc.com/news/health-31069173.

Garland, B., ed. 2004. *Neuroscience and the Law: Brain, Mind and the Scales of Justice*. New York: Dana Press.

Garland, B. and Frankel, M. S. 2006. Considering Convergence: A Policy Dialogue About Behavioral Genetics, Neuroscience, and Law. *Law and Contemporary Problems*, 69(1/2), pp. 101–113.

Gere, Cathy, and Parry, Bronwyn. 2006. The Flesh Made Word: Banking the Body in the Age of Information. *BioSocieties*, 1(1), pp. 41–54.

Gibbon, S. 2002. Re-examining Geneticization: Family Trees in Breast Cancer Genetics. *Science as Culture*, 11(4), pp. 429–457.

Gieryn, T. 1983. Boundary Work and the Demarcation of Science from Non-Science: Strains and Interests in Professional Ideologies of Scientists. *American Sociological Review*, 48(6), pp. 781–795.

Gonzalez-Polledo, E. J. and Posocco, S. 2020. Archives, Promises, Values: Forensic Infrastructures in Times of Austerity. *Critique of Anthropology*, 41(1), pp. 3–20.

Gonzalez-Polledo, E. J. and Posocco, S. 2021. Archives, promises, values: forensic infrastructures in times of austerity. *Critique of Anthropology*, 41(1), pp. 3–20.

Gonzalez-Polledo, Elena J. and Posocco, Silvia, eds. 2021. *Worlds and Futures*. New York and London: Routledge.

Greely, H. T. 2019. CRISPR'd Babies: Human Germline Genome Editing in the 'He Jiankui Affair'. *Journal of Law and the Biosciences*, 6(1), pp. 111–183.

Greely, H. T. and Farahany, N. A. 2019. Neuroscience and the Criminal Justice System. *Annual Review of Criminology*, 2, pp. 451–471.

Green, S., Carusi, A., and Hoeyer, K. 2019. Plastic Diagnostics: The Remaking of Disease and Evidence in Personalized Medicine. *Social Science & Medicine*, 304, 112318.

Greenhouse, Carol J. 2008. Life Stories, Law's Stories: Subjectivity and Responsibility in the Politicization of the Discourse of 'Identity.' *PoLAR: Political and Legal Anthropology Review*, 31(1) (May), pp. 79–95.

Greenhouse, C. J. 2014. Unexpected Properties: Strathern on the Relation of Law and Culture. *Theory, Culture & Society*, 31(2–3), pp. 167–184.

Guardian. 2020. 'There Is a Policy of Surrender': Doctor on UK's Covid-19 Failures, March 17. Available at https://www.theguardian.com/world/2020/mar/17/there-is-a-policy-of-surrender-doctor-on-uks-covid-19-failures.

Guardian. 2021. JCVI Failed to Back Youth Covid Jabs Despite Favourable Modelling, November 2. Available at https://www.theguardian.com/politics/2021/nov/02/jcvi-failed-back-youth-covid-jabs-despite-favourable-models.

Guthman, J. and Mansfield, B. 2013. The Implications of Environmental Epigenetics: A New Direction for Geographic Inquiry on Health, Space, and Nature-Society Relations. *Progress in Human Geography*, 37(4), pp. 486–504.

Habermas, J. 1995. Paradigms of Law. *Cardozo Law Review* 17, p. 771.

Habermas, J. 1996. *Between Facts and Norms*. Cambridge, MA: MIT Press.

Habermas, J. 2003. *The Future of Human Nature*. Cambridge: Polity Press.

Hacking, I. 2006a. Genetics, Biosocial Groups & The Future of Identity. *Daedalus*, 135(4), pp. 81–95.

Hacking, I. 2006b. Making up People. *London Review of Books*, 28(16) (August). https://www.lrb.co.uk/the-paper/v28/n16/ian-hacking/making-up-people.

Hamilton, J. A. 2008. Revitalizing difference in the HapMap: Race and contemporary human genetic variation research. *Journal of Law, Medicine & Ethics*, 36(3), pp.471–477.

Hamilton, Jennifer. 2012. The Case of the Genetic Ancestor. In Wailoo, K., Nelson, A., and Lee, C. (eds.), *Genetics and the Unsettled Past: The Collision of DNA, Race, and History*. New Brunswick: Rutgers University Press.

Hanafin, P. 2008. Voicing embodiment, relating difference: Towards a relational legal subjectivity. *Australian Feminist Law Journal*, 29(1), pp. 77–89.

Hanafin, P. 2013. Rights, Bioconstitutionalism and the Politics of Reproductive Citizenship in Italy. *Citizenship Studies*, 17(8), pp. 942–955.

Hanafin, P. 2014. Resistant Lives: Law, Life, Singularity. *Soft Power*, 1(1), pp. 91–107.

Haraway, Donna 1997. *Modest _Witness @Second_Millenium.FemaleMan_ Meets_OncoMouse™: Feminism and Technoscience*. New York: Routledge.

Haraway, Donna J. 2016. *Staying with the trouble: Making Kin in the Chthulucene*. Durham: Duke University Press.

Harris, C. 1993. Whiteness as Property. *Harvard Law Review*, 106(8), 1709.

Hart, H. L. 1961. *The Concept of Law*. Clarendon Law Series. Oxford: Clarendon Press.

Hayden, Cori. 2003. *When Nature Goes Public. The Making and Unmaking of Bioprospecting in Mexico*. Princeton and Oxford: Princeton University Press.

Hayden, C. 2005. Benefit-Sharing: Experiments in Governance. In Ghosh, R. (ed.), *CODE: Collaborative Ownership and the Digital Economy*. Cambridge: MIT Press.

Hayden, C. 2007. Taking as Giving: Bioscience, Exchange, and the Politics of Benefit Sharing. *Social Studies of Science*, 37(5), pp. 729–758.

Hefferon, K. and Herring, R. 2020. Engineered Gene Drives and Their Value in the Control of Vector-Borne Diseases, Weeds, Pests, and Invasive Species. In *GMOs*, pp. 401–419. Cham: Springer.

Heinemann, L. and Heinemann, T. 2010. Optimise your brain!–Popular science and its social implications. *BioSocieties*, 5, pp.291–294.

Heinemann, T. and Lemke, T. 2013. Suspect Families: DNA Kinship Testing in German Immigration Policy. *Sociology*, *47*(4), pp. 810–826.

Heinemann, T. and Lemke, T. 2014. Biological Citizenship Reconsidered: The Use of DNA Analysis by Immigration Authorities in Germany. *Science, Technology, & Human Values*, 39(4), pp. 488–510.

Helmreich, S. 2015. *Sounding the Limits of Life: Essays in the Anthropology of Biology and Beyond*. Princeton: Princeton University Press.

Hilgartner, Stephen. 2009. Intellectual Property and the Politics of Emerging Technology: Inventors, Citizens, and Powers to Shape to the Future. *Chicago-Kent Law Review*, 84(1), pp. 197–226.

Hill, John Lawrence. 1997. Law and the Concept of the Core Self: Toward a Reconciliation of Naturalism and Humanism. *Marquette Law Review*, 80, pp. 289–390.

Hinchcliffe, S. 2020. *Model Evidence—The COVID-19 Case*. https://somatosphere.com/forumpost/model-evidence-covid-19/.

Hinchliffe, S. Bingham, N. Allen, J. and Carter, S. 2016. *Pathological lives: Disease, space and biopolitics*. John Wiley & Sons.

Hinchliffe, S., Bingham, N., Allen, J., et al., eds. 2017. *Pathological Lives: Disease, Space and Biopolitics*. Malden, MA and Oxford, UK: Wiley Blackwell.

Hinterberger, A. 2012. Publics and populations: The politics of ancestry and exchange in genome science. *Science as Culture*, 21(4), pp. 528–549.

Hinterberger, A. and Porter, N., 2015. Genomic and viral sovereignty: Tethering the materials of global biomedicine. *Public Culture*, 27(2), pp.361–386.

Hird, M. J. 2007. The Corporeal Generosity of Maternity. *Body & Society*, 13(1), pp. 1–20.

Hobbes, T. 1996 [1651]. *Leviathan* (ed., R. Tuck). Cambridge: Cambridge University Press.

Hobsbawm, E. J. 1992. *Nations and Nationalism Since 1780: Programme, Myth, Reality*. Cambridge: Cambridge University Press.

Hodgson, D. 2000. Guilty mind or guilty brain? Criminal responsibility in the age of neuroscience. *Australian Law Journal*, 74(10), pp. 661–680.

Hoeyer, K. 2013. *Exchanging Human Bodily Material: Rethinking Bodies and Markets*. Dordrecht: Springer Science & Business Media.

Hoeyer, K. 2018. Lost and Found: Relocating the Individual in the Age of Intensified Data Sourcing in European Healthcare. *Personalized Medicine, Individual Choice and the Common Good*, pp. 144–166. Cambridge: Cambridge University Press.

Hoeyer, K. 2019. Data as Promise: Reconfiguring Danish Public Health Through Personalized Medicine. *Social Studies of Science*, 49(4), pp. 531–555.

Hoeyer, K. and Hogle, L.F. 2014. Informed Consent: the Politics of Intent and Practice in Medical Research Ethics. *Annual Review of Anthropology*, 43, pp. 347–362.

Holmes, O. W. 1899. Law in Science and Science in Law. *Harvard Law Review*, 12(7), pp. 443–463.

hooks, b. 2012. *Writing Beyond Race: Living Theory and Practice*. London: Routledge.

Howell, S. 2003. Kinning: The Creation of Life Trajectories in Transnational Adoptive Families. *Journal of the Royal Anthropological Institute*, 9(3), pp. 465–484.

Howell, S. 2009. Adoption of the Unrelated Child: Some Challenges to the Anthropological Study of Kinship. *Annual Review of Anthropology*, 38, pp. 149–166.

Huffman, Carl. 2021. *Alcmaeon: The Stanford Encyclopedia of Philosophy* (ed., N. Zalta). https://plato.stanford.edu/archives/sum2021/entries/alcmaeon/.

Hurlbut, J. B. 2020. Imperatives of Governance: Human Genome Editing and the Problem of Progress. *Perspectives in Biology and Medicine*, 63(1), pp. 177–194.

Hurlbut, J.B. 2017. A science that knows no country: Pandemic preparedness, global risk, sovereign science. *Big Data & Society*, 4(2), p. 1–14.

Ihar, Z. D. 2020. Phenotypic Personhood: Epigenetics and the Biolegality of Processing Asylum. In *Personhood in the Age of Biolegality*, pp. 127–147. Cham: Palgrave Macmillan.

Ingold, Tim and Palsson, Gisli, eds. 2013. *Biosocial Becomings: Integrating Social and Biological Anthropology*. Cambridge, UK: Cambridge University Press.

Isin, E. and Ruppert, E. 2020. The Birth of Sensory Power: How a Pandemic Made it Visible?. *Big Data & Society*, 7(2), pp. 1–20.

Jackson, E. 2018. The Ambiguities of 'Social' Egg Freezing and the Challenges of Informed Consent. *BioSocieties*, 13(1), pp. 21–40.

Jasanoff, S. 1997. *Science at the Bar: Law, Science, and Technology in America*, vol. 9. Cambridge: Harvard University Press.

Jasanoff, S. 2001. Ordering Life: Law and the Normalization of Biotechnology. *Politeia*, 62, pp. 34–50.

Jasanoff, S., ed. 2004. *States of Knowledge: The Co-production of Science and the Social Order*. London: Routledge.

Jasanoff, S. 2005. *Designs on Nature: Science and Democracy in Europe and the United States*. Princeton: Princeton University Press.

Jasanoff, S. 2006. Just Evidence: The Limits of Science in the Legal Process. *The Journal of Law, Medicine & Ethics*, 34(2), pp. 328–341.

Jasanoff, S. 2007. Technologies of Humility. *Nature*, 450(7166), p. 33.

Jasanoff, S. 2010. Foreword. In Hindmarsh, R. and Prainsack, B. (eds.), *Genetic Suspects: Global Governance of Forensic DNA Profiling and Databasing*. Cambridge: Cambridge University Press.

Jasanoff, S. 2012. *Science and Public Reason*. Routledge.

Jasanoff, S., ed. 2011. *Reframing Rights: Bioconstitutionalism in the Genetic Age*. Cambridge: MIT Press.

Jasanoff, Sheila. 2015. Serviceable Truths: Science for Action in Law and Policy. *Texas Law Review*, 93(7), pp. 1723–1749.

Jasanoff, S. 2016. Perfecting the Human: Posthuman Imaginaries and Technologies of Reason. In *Perfecting Human Futures*, pp. 73–95. Wiesbaden: Springer VS.

Jasanoff, S. 2020. Pathologies of Liberty: Public Health Sovereignty and the Political Subject in the Covid-19 Crisis. *Cahiers Droit, Sciences & Technologies*, 11, pp. 125–149.

Jasanoff, S. and Kim, S. H. 2015. *Dreamscapes of Modernity: Sociotechnical Imaginaries and the Fabrication of Power*. Chicago: University of Chicago Press.

Jasanoff, S., Hurlbut, J.B., and Saha, K. 2015. CRISPR Democracy: Gene Editing and the Need for Inclusive Deliberation. *Issues in Science and Technology*, 32(1), pp. 25–32.

Joly, Yann, Braker, Maria, and Le Huynh, Michael. 2010. Genetic Discrimination in Private Insurance: Global Perspectives. *New Genetics and Society*, 29(4), pp. 351–368.

Jones, D. S. 2013. How Personalized Medicine Became Genetic, and Racial: Werner Kalow and the Formations of Pharmacogenetics. *Journal of the History of Medicine and Allied Sciences*, 68, pp. 1–48.

Jones, David S. and Helmreich, Stefan. 2020. The Shape of Epidemics. *Boston Review*, June 26. https://www.bostonreview.net/articles/david-shumway-jones-stefan-helmreich-epidemic-waves/.

Kahn, J. 2004. How a Drug Becomes 'Ethnic': Law, Commerce, and the Production of Racial Categories in Medicine. *Yale Journal of Health Policy, Law, and Ethics*, 4(1), pp. 1–46.

Kahn, J. 2006. Genes, Race, and Population: Avoiding a Collision of Categories. *American Journal of Public Health*, 96(11), pp. 1965–1970.

Kahn, J. 2015. Neuroscience, Sincerity, and the Law. *Bergen Journal of Criminal Law & Criminal Justice*, 3(2), pp. 203–220.

Kahn, J. 2022. Anthropology, Law, and the Problem of Incommensurability. *Alabama Law Review*, 73(4), pp. 783–801.

Kang, H. Y. 2012. 'Science Inside Law: The Making of a New Patent Class in the International Patent Classification', *Science in Context*, 25(4), 551–194.

Kang, H.Y., 2018. Law's Materiality: Between Concrete Matters and Abstract Forms, or How Matter Becomes Material. In *Routledge Handbook of Law and Theory*, pp. 453–474. Routledge.

Kang, H. Y. 2020. Patents as Assets: Intellectual Property Rights as Market Subjects and Objects. In Birch, K. and Muniesa, F. (eds.), *Assetization: Turning Things into Assets in Technoscientific Capitalism*, pp. 45–74. Cambridge: MIT Press.

Kang, H.Y. and Kendall, S. 2019. Legal Materiality. In S. Stern, M. DelMar, and B. Meyler (eds.), *The Oxford Handbook of Law and Humanities*, pp. 21–38. Oxford, UK: Oxford Univ. Press

Karpin, I. 2016. Regulatory responses to the gendering of transgenerational harm. *Australian Feminist Studies*, 31(88), pp. 139–153.

Keaney, J., 2022. The racializing womb: Surrogacy and epigenetic kinship. *Science, Technology, & Human Values, 47*(6), pp.1157–1179.

Keller, E. F. 2015. The Postgenomic Genome. In Sarah S. and Stevens, H. (eds.), *Postgenomics: Perspectives on Biology After the Genome*, pp. 9–31. Durham: Duke University Press.

Kelley, D. R. 1990. *The Human Measure: Social Thought in the Western Legal Tradition.* Cambridge: Harvard University Press.

Kemp, P. 1998. The Bioethical Turn. In Rendtorff, Jacob and Kemp, Peter (eds.), *From Ethics to Biolaw*, pp. 9–20. Copenhagen: Centre for Ethics and Law, University of Copenhagen.

Kemp, P. 2000. Bioethics and Law and Biolaw in Ethics. In Kemp, Peterm, Rendtorff, Jacob, and Johansen, Niels Mattson (eds.), *Bioethics and Biolaw*, vol. 1, *Judgement of Life*, pp. 63–77. Copenhagen: Rhodos International Science and Art Publishers and Centre for Ethics and Law.

Kirksey, E., ed. 2014. *The Multispecies Salon.* Durham: Duke University Press.

Kowal, E., Radin, J. and Reardon, J. 2013. Indigenous body parts, mutating temporalities, and the half-lives of postcolonial technoscience. *Social Studies of Science*, 43(4), pp.465–483.

Kowal, E. and Radin, J. 2015. Indigenous Biospecimen Collections and the Cryopolitics of Frozen Life. *Journal of Sociology*, 51(1), pp. 63–80.

Kowal, E. and Watt, E. 2018. What is race in Australia. *Journal of Anthropological Sciences*, 96, pp.229–237.

Kroløkke, C. H. and Adrian, S. W. 2013. Sperm on Ice: Fatherhood and Life After Death. *Australian Feminist Studies*, 28(77), pp. 263–278.

Kudlek, Karolina. 2022. Towards a Systematic Evaluation of Moral Bioenhancement. *Theoretical Medical Bioethics*, 43(2–3), pp. 95–110.

Lafuente-Funes, S. 2023. The Role of Vitrification in Spanish Reproductive Labs: A Cryo-revolution Led by Strategic Freezing. *Science, Technology, & Human Values*, 48(4), pp. 752–776.

Lakoff, Andrew. 2017. *Unprepared: Global Health in a Time of Emergency*. Oakland: University of California Press.

Landecker, H. 1999. Between Beneficence and Chattel: the Human Biological in Law and Science. *Science in Context*, 12(1), pp. 203–225.

Landecker, H. 2005. Living Differently in Time: Plasticity, Temporality and Cellular Biotechnologies. *Culture Machine* 7. http://www.culturemachine.net/index.php/cm/article/view/26/33.

Landecker, H. and Panofsky, A. 2013. From Social Structure to Gene Regulation, and Back: A Critical Introduction to Environmental Epigenetics for Sociology. *Annual Review of Sociology*, 39, pp. 333–357.

Lappé, M. 2016. The maternal body as environment in autism science. *Social Studies of Science*, 46(5), pp.675–700.

Lappé, M. Jeffries Hein, R. and Landecker, H. 2019. Environmental politics of reproduction. *Annual Review of Anthropology*, 48, pp.133–150.

Lappé, M. and Landecker, H. 2015. How the genome got a life span. *New genetics and society*, 34(2), pp.152–176.

Latour, B. 1987. *Science in Action: How to Follow Scientists and Engineers Through Society*. Cambridge, MA: Harvard University Press.

Latour, B. 1999. *Pandora's Hope: Essays on the Reality of Science Studies*. Cambridge: Harvard University Press.

Latour, B. 2004. Scientific Objects and Legal Objectivity. *Law, Anthropology, and the Constitution of the Social: Making Persons and Things*, pp. 73–114. Cambridge: Cambridge University Press.

Latour, B. 2010. *The making of law: an ethnography of the Conseil d'État*. Cambridge: Polity.

Latour, B. 2012. *We Have Never Been Modern*. Cambridge: Harvard University Press.

Latour, B. 2013. *An Inquiry into Modes of Existence*. Harvard University Press.

Latour, B. 2021. Is This a Dress Rehearsal? *Critical Inquiry*, 47(S2), S25–S27.

Lawson, C. and Rourke, M. 2016. Open Access DNA, RNA and Amino Acid Sequences: The Consequences and Solutions for the International Regulation of Access and Benefit Sharing. *Griffith University Law School Research Paper* (16-12).

Lebner, A. 2000. Genetic "Mysteries" and International Adoption: The Cultural Impact of Biomedical Technologies on the Adoptive Family Experience. *Family Relations*, 49(4), pp. 371–377.

Lee, C., 2012. The Unspoken Significance of Gender in Constructing Kinship, Race, and Nation. In Wailoo, K., Nelson, A. And Lee, C. Eds., Genetics and the Unsettled Past: the Collision of Dna, Race, and History. *Rutgers University Press*. pp. 32–40.

Leeds, J. P. 1991. Moore v. Regents of the University of California: More for Biotechnology, Less for Patients. *Indiana Law Review*, 25(2), pp. 559–588.

Lemke, T. 2010. From State Biology to the Government of Life: Historical Dimensions and Contemporary Perspectives of 'Biopolitics'. *Journal of Classical Sociology*, 10(4), pp. 421–438.

Lemke, T. 2011. Biopolitics. In *Biopolitics*. New York University Press.

Lemke, T., Casper, M., and Moore, L. J. 2011. *Biopolitics: An Advanced Introduction*, vol. 5. New York: NYU Press.

Leonelli, S., 2010. The commodification of knowledge exchange: Governing the circulation of biological data. in Radder, H. (Ed.) *The commodification of academic research: Science and the modern university*. University of Pittsburgh Press. pp. 132–157.

Leonelli, Sabina. 2015. What Counts as Scientific Data? A Relational Framework. *Philosophy of Science*, 82(5), pp. 810–821.

Leonelli, S. and Ankeny, R.A. 2012. Re-thinking Organisms: The Impact of Databases on Model Organism Biology. *Studies in History and Philosophy of Biological and Biomedical Sciences* 43(1), 29–36.

Levine, I. T. 2022. Modeling Sovereignty: Towards a Legal Philosophy for Platforms in the People's Republic of China. *Peking University Law Journal*, 10(1), pp. 25–73.

Lezaun, J. 2006. Creating a New Object of Government: Making Genetically Modified Organisms Traceable. *Social Studies of Science*, 36(4), pp. 499–531.

Lichtenstein, P., Halldner, L., and Zetterqvist, J. 2012. Medication for Attention Deficit-Hyperactivity Disorder and Criminality. *New England Journal of Medicine*, 367, pp. 2006–2014.

Lock, M. 2013. The Epigenome and Nature/Nurture Reunification: A Challenge for Anthropology. *Medical Anthropology*, 32(4), pp. 291–308.

Lock, M. M. and Palsson, G. 2016. *Can Science Resolve the Nature/Nurture Debate?* New York: Wiley.

Locke, John. 1960 [1698]. *Two Treatises on Government*. Cambridge: Cambridge University Press.

Lockwood, G. M. 2011. Social EGG Freezing: The Prospect of Reproductive 'Immortality' or a Dangerous Delusion? *Reproductive Biomedicine Online*, 23(3), pp. 334–340.

Lorimer, J. 2019. Hookworms Make us Human: The Microbiome, Eco-Immunology, and a Probiotic Turn in Western Health Care. *Medical Anthropology Quarterly*, 33(1), pp. 60–79.

Lorimer, J. 2020. *The Probiotic Planet: Using Life to Manage Life*. Minneapolis: University of Minnesota Press.

Luhmann, N. 1988. Law as a Social System. *Northwestern University Law Review*, *83*, p. 136.

Luhmann, N. 2004. *Law as a Social System*. Oxford: Oxford University Press on Demand.

Lynch, M. 2003. God's Signature: DNA Profiling, the New Gold Standard in Forensic Science. *Endeavour*, 27(2), pp. 93–97.

Lynch, M. and McNally, R. 2009. Forensic DNA Databases and Biolegality. *Handbook of Genetics and Society*, pp. 283–301. London: Routledge.

Lynch, M., Cole, S. A., McNally, R., and Jordan, K. 2010. *Truth Machine: The Contentious History of DNA Fingerprinting*. Chicago: University of Chicago Press.

Mackenzie, Adrian and Roberts, Celia. 2017. Adopting Neuroscience: Parenting and Affective Indeterminacy. *Body & Society*, 23(3), pp. 130–155.

MacKinnon, Catherine A. 1989. Sexuality, Pornography, and Method: Pleasure Under Patriarchy. *Ethics*, 99(2), pp. 314–346.

Manderson, L. 2016. Foetal politics and the prevention of chronic disease. *Australian Feminist Studies*, 31(88), pp.154–171.

Mansfield, B. 2012. Race and the New Epigenetic Biopolitics of Environmental Health. *BioSocieties*, 7, pp. 352–372.

Mansfield, B. and Guthman, J. 2015. Epigenetic Life: Biological Plasticity, Abnormality, and New Configurations of Race and Reproduction. *Cultural Geographies*, 22(1), pp. 3–20.

Margaret, D. 2023. *Asking the Law Question*, 5th ed. Sydney: Lawbook Co.

Margulies, D. S. 2012. The Salmon of Doubt: Six Months of Methodological Controversy Within Social Neuroscience. In Choudhury, S. and Slaby, J. (eds.), *Critical Neuroscience: A Handbook of the Social and Cultural Contexts of Neuroscience*, pp. 273–285. Malden: Wiley-Blackwell.

Mascolo, J. and Bui, J. 2023. Chemical and Surgical Castration for Sex Offenders. *FindLaw*, https://www.findlaw.com/criminal/criminal-charges/chemical-and-surgical-castration.html

Maurer, B. 2004. The Cultural Power of Law? Conjunctive Readings. *Law & Society Review*, 38(4), pp. 843–850.

Mbembe, A. 2008. Necropolitics. In *Foucault in an Age of Terror: Essays on Biopolitics and the Defence of Society*, pp. 152–182. London: Palgrave Macmillan UK.

M'charek, A. 2008. Silent Witness, Articulate Collective: DNA Evidence and the Inference of Visible Traits. *Bioethics*, 22(9), pp. 519–528.

M'charek, A. 2021. Traces of Race, Roots of Gender. *The Routledge Companion to Black Women's Cultural Histories*. London: Routledge.

M'charek, A. and Casartelli, S. 2019. Identifying Dead Migrants: Forensic Care Work and Relational Citizenship. *Citizenship Studies*, 23(7), pp. 738–757.

M'charek, A., Schramm, K., and Skinner, D. 2014. Topologies of Race: Doing Territory, Population and Identity in Europe. *Science, Technology, & Human Values*, 39(4), 468–487.

M'charek, A., Toom, V., and Jong, L. 2020. The Trouble with Race in Forensic Identification. *Science, Technology, & Human Values*, 45(5), pp. 804–828.

M'charek, A., Toom, V. and Prainsack, B. 2012. Bracketing off population does not advance ethical reflection on EVCs: A reply to Kayser and Schneider. *Forensic Science International: Genetics*, 6(1), pp.e16–e17.

Melhuus, M. 2009. Conflicting Notions of Continuity and Belonging: Assisted Reproduction, Law, and Practices in Norway. *Social Analysis*, 53(3), pp. 148–162.

Meloni, M. 2015. Epigenetics for the Social Sciences: Justice, Embodiment and Inheritance in the Postgenomic Age. *New Genetics and Society*, 34(2), pp. 125–151.

Meloni, M. 2016. *Political Biology: Science and Social Values in Human Heredity from Eugenics to Epigenetics*, p. 295. Basingstoke: Palgrave Macmillan.

Meloni, M. 2017. Race in an Epigenetic Time: Thinking Biology in the Plural. *British Journal of Sociology*, 68(3), pp. 389–409.

Meloni, M. 2018. A Postgenomic Body: Histories, Genealogy, Politics. *Body & Society*, 24(3), pp. 3–38.

Meloni, M. and Testa, G. 2014. Scrutinizing the Epigenetics Revolution. *BioSocieties*, 9, pp. 431–456.

Mills, C. 2011. *Futures of Reproduction: Bioethics and Biopolitics*. New York: Springer Science & Business Media.

Mills, C. 2017. *Biopolitics*. London: Routledge.

Mirowski, P. 2011. *Science-mart: Privatizing American Science*. Cambridge: Harvard University Press.

Mitchell, R. and Waldby, C. 2006. *Tissue Economies: Blood, Organs, and Cell lines in Late Capitalism*. Durham: Duke University Press.

Mitchell, R. and Waldby, C. 2010. National biobanks: clinical labor, risk production, and the creation of biovalue. *Science, Technology, & Human Values*. 35(3), pp.330–355.

Mitra, S., Schicktanz, S., and Patel, T., ed. 2018. *Cross-Cultural Comparisons on Surrogacy and Egg Donation: Interdisciplinary Perspectives from India, Germany and Israel*. Cham: Palgrave Macmillan.

Modell, J. 2002. *A Sealed and Secret Kinship: The Culture of Policies and Practices in American Adoption*. New York: Berghahn Books.

Montoya, M. J. 2007. Bioethnic conscription: Genes, race, and Mexicana/o ethnicity in diabetes research. *Cultural anthropology*, 22(1), pp. 94–128.

Morse, S. J. 2006. Moral and legal responsibility and the new neuroscience. Neuroethics: Defining the issues in theory, *practice, and policy*, pp. 33–50.

Morse, S. J. 2011a. Neuroscience and the Future of Personhood and Responsibility. In J. Rosen and B. Wittes (eds.), *Constitution 3.0: Freedom and Technological Change*, pp. 113–129. Washington, DC: Brookings Institution Press.

Morse, Stephen J. 2011b. Lost in Translation? An Essay on Law and Neuroscience. *Law and Neuroscience,* 13, pp. 529–562.

Morse, Stephen J. 2014. Neuroethics: Neurolaw. *The Oxford Handbook of Topics in Philosophy* (online ed., Oxford Academic, April 1). https://doi.org/10.1093/oxfordhb/9780199935314.013.45. Accessed 6 August 2023.

Morse, Stephen J. 2016. Actions speak louder than images: The use of neuroscientific evidence in criminal cases. *Journal of Law and the Biosciences*, 3(2), pp.336–342.

Morse, Stephen J. 2017. Neuroethics: Neurolaw. *The Oxford Handbook of Topics in Philosophy*. https://doi.org/10.1093/oxfordhb/9780199935314.013.45. Accessed 6 Mar. 2024.

Murphy, H. 2018. What 13,000 Patents Involving the DNA of Sea Life Tell Us About the Future. *The New York Times*, September 17. https://www.nytimes.com/2018/09/17/science/patents-marine-dna.html.

Mutcherson, K. M. 2016. Procreative Rights in a Postcoital World. *The Oxford Handbook of Reproductive Ethics*, pp. 159–181. New York: Oxford University Press.

Naffine, N. 2009. *Law's Meaning of Life: Philosophy, Religion, Darwin and the Legal Person*. London: Bloomsbury Publishing.

Nash, Catherine. 2002. Genealogical Identities. *Environment and Planning D: Society and Space*, 20(1), pp. 27–52.

Nash, Catherine. 2004. Genetic Kinship. *Cultural Studies*, 18(1), pp. 1–33.

Nash, C. 2005. Geographies of relatedness. *Transactions of the Institute of British Geographers*, 30(4), pp. 449–462.

Nelkin D, Lindee S. 1995. *The DNA Mystique: The Gene as a Cultural Icon.* New York: Freeman

Nelson, A. 2008. Bio science: Genetic genealogy testing and the pursuit of African ancestry. *Social studies of science*, 38(5), pp.759–783.

Nelson, A. 2016. *The Social Life of DNA: Race, Reparations, and Reconciliation after the Genome*, Boston, MA: Beacon Press.

Nguyen, V. 2020. Breathless in Beijing: Aerial Attunements and China's New Respiratory Publics. *Engaging Science, Technology, and Society*, 6, pp. 439–461.

Nicol, D. Bubela, T. Chalmers, D. Charbonneau, J. Critchley, C. Dickinson, J. Fleming, J. Hewitt, A. W. Kaye, J. Liddicoat, J. and McWhirter, R., 2016. Precision medicine: drowning in a regulatory soup?. *Journal of Law and the Biosciences*, 3(2), pp. 281–303.

Niewöhner, J. 2011. Epigenetics: Embedded Bodies and the Molecularisation of Biography and Milieu. *BioSocieties*, 6(3), pp. 279–298.

Novas, C. and Rose, N. 2000. Genetic Risk and the Birth of the Somatic Individual. *Economy and Society*, 29(4), pp. 485–513.

Nowotny, H. Scott, P. and Gibbons, M. 2001. *Re-thinking science: Knowledge and the public in an age of uncertainty* (p. 12). Cambridge: Polity.

Obasogie, O. K. 2008. Beyond best practices: strict scrutiny as a regulatory model for race-specific medicines. *Journal of Law, Medicine & Ethics*, 36(3), pp.491–497.

O'Malley, M. A. and Skillings, D. J. 2018. Methodological Strategies in Microbiome Research and Their Explanatory Implications. *Perspectives on Science*, 26(2), pp. 239–265.

Obasogie, O. K. and Darnovsky, M., eds. 2018. *Beyond Bioethics: Toward a New Biopolitics*. Berkeley: University of California Press.

O'Connor, Cliodhna, Rees, Geraint, and Joffe, Helene. 2012. Neuroscience in the Public Sphere. *Neuron*, 74(2), pp. 220–226.

O'Malley, M. A. and Skillings, D. J. 2018. Methodological strategies in microbiome research and their explanatory implications. *Perspectives on Science*, 26(2), pp.239–265.

Ong, A. 2010 'An Analytics of Biotechnology and Ethics at Multiple Scales', in Ong and Nancy N. Chen, eds, *Asian Biotech: Ethics and Communities of Fate* (Durham NC: Duke University Press, 2010).

Ong, A. 2016. *Fungible life: Experiment in the Asian city of life*. Duke University Press.

Ossorio, Pilar and Duster, Troy. 2005. Race and Genetics: Controversies in Biomedical, Behavioral, and Forensic Sciences. *American Psychologist*, 60(1), p. 115.

O'Conner, C., Rees, G., and Joffe, H. 2012. Neuroscience in the public sphere. *Neuron*, 74, 220–226.

Oudshoorn, Nelly. 2003. *Beyond the Natural Body: An Archaeology of Sex Hormones*. London and New York: Routledge.

Ozdemir, V. Muljono, D. H. Pang, T. Ferguson, L. R. Manamperi, A., Samper, S. Someya, T. Tassé, A. M. Tsai, S. J. Zhou, H. H. and Lee, E. J. 2011. Asia-Pacific Health 2020 and genomics without borders: co-production of knowledge by science and society partnership for global personalized medicine. *Current Pharmacogenomics and Personalized Medicine*, 9(1), p.1.

Palmer, C. 2016. Human and Object, Subject and Thing: The Troublesome Nature of Human Biological Material (HBM). *Contested Categories: Life Sciences in Society*, pp. 15–30. London: Routledge.

Petersen, Alan. 2010. *The Politics of Bioethics*. New York: Routledge.

Parry, Bronwyn. 2004. *Trading the Genome: Investigating the Commodification of Bio-information*. New York: Columbia University Press.

Parry, Bronwyn and Gere, Cathy. 2006. Contested Bodies: Property Models and the Commodification of Human Biological Artefacts. *Science as Culture*, 15(2), pp. 139–158.

Parry, Bronwyn and Greenhough, Beth. 2017. *Bioinformation*. New York: Wiley.

Parthasarathy, S. 2011. Whose Knowledge? What Values? The Comparative Politics of Patenting Life Forms in the United States and Europe. *Policy Sciences*, 44, pp. 267–288.

Parthasarathy, S. 2017. *Patent Politics: Life Forms, Markets, and the Public Interest in the United States and Europe*. University of Chicago Press.

Pavlich, G. 2016. Avowal and criminal accusation. *Law and Critique*, 27(2), pp.229–245.

Pavone, V. and Goven, J., eds. 2017. *Bioeconomies*. Cham: Palgrave Macmillan.

Paxson, H. 2008. Post-pasteurian cultures: The microbiopolitics of raw-milk cheese in the United States. *Cultural anthropology*, 23(1), pp.15–47.

Pernu, T. K. and Elzein, N. 2020. From Neuroscience to Law: Bridging the Gap. *Frontiers in Psychology*, 11, p. 1862.

Petersen, A. 2011. *The Politics of Bioethics*. Routledge.

Petryna, A. 2004. Biological Citizenship: The Science and Politics of Chernobyl-exposed Populations. *Osiris*, 19, pp. 250–265.

Petryna, A. 2013. *Life Exposed: Biological Citizens after Chernobyl*. Princeton University Press.

Pickersgill, M. 2009. Between Soma and Society: Neuroscience and the Ontology of Psychopathy. *BioSocieties*, 4(1), pp. 45–60.

Pickersgill, M. 2011. Connecting Neuroscience and Law: Anticipatory Discourse and the Role of Sociotechnical Imaginaries. *New Genetics and Society*, 30(1), pp. 27–40.

Pickersgill, M. 2013. The Social Life of the Brain: Neuroscience in Society. *Current Sociology*, 61(3), pp. 322–340.

Pickersgill, M. 2013. How Personality Became Treatable: The Mutual Constitution of Clinical Knowledge and Mental Health Law. *Social Studies of Science*, 43(1), pp. 30–53.

Pickersgill, M., Niewöhner, J., Müller, R., Martin, P. and Cunningham-Burley, S. 2013. Mapping the New Molecular Landscape: Social Dimensions of Epigenetics. *New Genetics and Society*, 32(4), pp. 429–447.

Pitts-Taylor, V. 2016. *The Brain's Body: Neuroscience and Corporeal Politics*. Durham, NC: Duke University Press.

Poland, S. C. 2005. Bioethics, Biolaw, and Western Legal Heritage. *Kennedy Institute of Ethics Journal*, 15(2), pp. 211–218.

Pollock, Anne. 2008. Pharmaceutical Meaning-making Beyond Marketing: Racialized Subjects of Generic Thiazide. *Journal of Law, Medicine & Ethics*, 36(3), pp. 530–536.

Pollock, A. 2019. *Synthesizing Hope: Matter, Knowledge, and Place in South African Drug Discovery*. University of Chicago Press.

Posner, Richard A. 1983. *The Economics of Justice*. Cambridge: Harvard University Press.

Pottage, A. 1998. The Inscription of Life in Law: Genes, Patents, and Biopolitics. *Modern Law Review*, 61, p. 740.

Pottage, A. 2002. Unitas personae: on legal and biological self-narration. *Law & Literature*, 14(2), pp.275–308.

Pottage, A. 2004. Introduction: The Fabrication of Persons and Things. In Pottage, A. (ed.), *Law, Anthropology, and the Constitution of the Social: Making Persons and Things*. Cambridge: Cambridge University Press.

Pottage, A. 2007. The Socio-legal Implications of the New Biotechnologies. *Annual Review of Law and Social Science*, 3, pp. 321–344.

Pottage, A. 2012. The Materiality of What? *Journal of Law and Society*, 39(1), pp. 167–183.

Pottage, A. Mundy, M. 2004. eds. Law, Anthropology, and the Constitution of the Social: Making Persons and Things. Cambridge Studies in Law and Society. Cambridge University Press.

Pottage, A. and Sherman, B. 2010. *Figures of Invention: A History of Modern Patent Law*. Oxford: Oxford University Press.

Prainsack, B. 2010. Key issues in DNA profiling and databasing: Implications for governance. In Hindmarsh, R. and Prainsack, B. eds., 2010. *Genetic suspects: Global governance of forensic DNA profiling and databasing*. pp.153–174, Cambridge: Cambridge University Press.

Prainsack, B. 2017. Personalized medicine: empowered patients in the 21st century?. In Personalized Medicine. New York University Press.

Prasad, A. 2005. Making Images/Making Bodies: Visibilizing and Disciplining Through Magnetic Resonance Imaging (MRI). *Science, Technology & Human Values*, 30(2), pp. 291–316.

Pugliese, J. 2012. *Biometrics: Bodies, Technologies, Biopolitics*. London: Routledge.

Purvis, D. E. 2012. Intended Parents and the Problem of Perspective. *Yale Journal of Law and Feminism*, 24, p. 210.

Quigley, M. 2018. *Self-ownership, Property Rights, and the Human Body: a Legal and Philosophical Analysis (Vol. 43)*. Cambridge University Press.

Rabinow. P. 1992. Artificiality and the Enlightenment: From Sociobiology to Biosociality. In Crary, Jonathan and Kwinter, Sanford (eds.), *Incorporations*. New York: Zone Books.

Rabinow, P. 1996. Artificiality and Enlightenment: From Sociobiology to Biosociality. In *Essays on the Anthropology of Reason*, pp. 91–111. Princeton, NJ: Princeton University Press.

Rabinow, P. 1999. Artificiality and Enlightenment: From Sociobiology to Biosociality. In *Essays on the Anthropology of Reason*. Princeton: Princeton University Press.

Rabinow, P. 2005. Artificiality and Enlightenment: From Sociobiology to Biosociality. In Inda, Jonathan Xavier and Rabinow, Paul (eds.), *Anthropologies of Modernity: Foucault, Governmentality, and Life Politics*, pp. 179–193. Malden: Blackwell Publishing.

Rabinow, P. 2010. Artificiality and Enlightenment: From Sociobiology to Biosociality. *Politix*, 2, pp. 21–46.

Rabinow, P., 2011. *Making PCR: A story of biotechnology*. University of Chicago Press.

Rabinow, P. and Bennett, G. 2012. *Designing Human Practices: An Experiment with Synthetic Biology*. Chicago: University of Chicago Press.

Rabinow, P. and Rose, N. 2006. Biopower today. *BioSocieties*, 1(2), pp. 195–217.

Radin, J. 2018. Ethics in Human Biology: a Historical Perspective on Present Challenges. *Annual Review of Anthropology*, 47, pp. 263–278.

Rajagopalan, R. M., Nelson, A., and Fujimura, J. H. 2017. 12 Race and Science in the Twenty-First Century. In Felt, U., Fouché, R., Miller, C. A., and Smith-Doerr, L. (eds.), *The Handbook of Science and Technology Studies*, 4th ed., pp. 349–378. Cambridge: MIT Press.

Rajan, K. S. 2006. *Biocapital: The Constitution of Postgenomic Life*. Durham: Duke University Press.

Rajan, K.S. 2017. *Pharmocracy: Value, politics, and knowledge in global biomedicine*. Duke University Press.

Rajan, Kaushik Sunder and Leonelli, Sabina. 2013. Introduction: Biomedical trans-actions, Postgenomics, and Knowledge/Value. *Public Culture*, 25(3), pp. 463–475.

Rajagopalan, Ramya M., Nelson, Alondra, and Fujimura, Joan H. 2017. Race and Science in the Twenty-First Century. In Felt, U., Fouche, R., Miller, C. A., and Smith-Doerr, L. (eds.), *The Handbook of Science and Technology Studies*, 4th ed., pp. 349–378. Cambridge, MA: The MIT Press.

Raffles, H. 2001. The Uses of Butterflies. *American Ethnologist*, 28(3), pp. 513–548

Rafter, N. H. 1997. *Creating Born Criminals*. Urbana: University of Illinois Press.

Rafter, N.H. ed., 2009. *The origins of criminology: A reader*. London: Routledge

Raman, Sujatha and Tutton, Richard. 2010. Life, Science, and Biopower. *Science, Technology, & Human Values*, 35(5), pp. 711–734.

Rao, R. 1995. Assisted Reproductive Technology and the Threat to the Traditional Family. *Hastings Law Journal*, 47, p. 951.

Reardon, J. 2009. *Race to the Finish: Identity and Governance in an Age of Genomics*. Princeton University Press.

Reardon, Jenny. 2017. *The Postgenomic Condition: Ethics, Justice, and Knowledge After the Genome*. Chicago, IL: University of Chicago Press.

Reardon, Jenny and TallBear, Kim. 2012. "Your DNA Is Our History" Genomics, Anthropology, and the Construction of Whiteness as Property. *Current Anthropology*, 53(S5), S233–S245.

Rees, T. 2016. *Plastic reason: An anthropology of brain science in embryogenetic terms*. University of California Press.

Reilly, P. R. 2015. Eugenics and Involuntary Sterilization: 1907–2015. *Annual Review of Genomics in Human Genetics*, Vol. 16, 351–368.

Rhodes, T. and Lancaster, K. 2020. Mathematical Models as Public Troubles in COVID-19 Infection Control: Following the Numbers. *Health Sociology Review*, 29(2), pp. 177–194.

Rhodes, T., Lancaster, K., and Rosengarten, M. 2020. A Model Society: Maths, Models and Expertise in Viral Outbreaks. *Critical Public Health*, 30(3), pp. 253–256.

Richardson, S. 2015. Maternal Bodies in the Postgenomic Order: Gender and the Explanatory Landscape of Epigenetics. In Richardson, Sarah S. and Stevens, H. (eds.), *Postgenomics: Perspectives on Biology After the Genome*, pp. 210–231. Durham: Duke University Press.

Richardson, S.S. and Stevens, H., eds. 2015. *Postgenomics: Perspectives on Biology after the Genome*. Duke University Press.

Riles, A. 2005. A New Agenda for the Cultural Study of Law: Taking on the Technicalities. *Buffalo Law Review*, 53, p. 973.

Riles, Annelise. 2008. The Anti-network: Private Global Governance, Legal Knowledge, and the Legitimacy of the State. *The American Journal of Comparative Law*, 56(3), pp. 605–630.

Roberts, Celia. 2002. A Matter of Embodied Fact Sex Hormones and the History of Bodies. *Feminist Theory*, 3(1), pp. 7–26.

Roberts, D. E. 1995. Social Justice, Procreative Liberty, and the Limits of Liberal and the Limits of Liberal Theory: Robertson's Children of Choice. *Law & Social Inquiry*, 20(4), pp. 1005–1021.

Roberts, D. E. 2006. Legal Constraints on the Use of Biomedical Research: Toward a Social Justice Framework. *Journal of Law, Medicine & Ethics*, 24(3), pp. 526–534.

Roberts, D. E. 2008. Is Race-based Medicine Good for Us? African American Approaches to Race, Biomedicine, and Equality. *Journal of Law, Medicine & Ethics*, 36(3), pp. 537–545.

Roberts, D. E. 2013. Law, race, and biotechnology: toward a biopolitical and transdisciplinary paradigm. *Annual Review of Law and Social Science*, 9, pp. 149–166.

Robertson, John A. 1995. Liberalism and the Limits of Procreative Liberty: A Response to My Critics. *Washington and Lee Law Review*, 52, p. 233.

Rose, N. 2001. The Politics of Life Itself. *Theory, Culture & Society*, 18(6), pp. 1–30.

Rose, N. 2006. *The Politics of Life Itself: Biomedicine, Power, and Subjectivity in the Twenty-First Century*. Princeton: Princeton University Press.

Rose, N. 2007. Molecular Biopolitics, Somatic Ethics and the Spirit of Biocapital. *Social Theory & Health*, 5(1), pp. 3–29.

Rose, N. 2010. 'Screen and Intervene': Governing Risky Brains. *History of the Human Sciences*, 23(1), pp. 79–105.

Rose, N. and Abi-Rached, J. M. 2013. *Neuro: The New Brain Sciences and the Management of the Mind*. Princeton, NJ: Princeton University Press.

Rose, N. and Abi-Rached, J. M. 2014. Governing Through the Brain: Neuropolitics, Neuroscience and Subjectivity. *The Cambridge Journal of Anthropology*, 32(1), pp. 3–23.

Rose, N. and Novas, C. 2005. Biological Citizenship. *Global Assemblages: Technology, Politics, and Ethics as Anthropological Problems*, pp. 439–463. Oxford: Blackwell Publishing.

Rosenberg, C. E. 1993. *Explaining Epidemics*. Cambridge: Cambridge University Press.

Roy, V. 2020. A Crisis for Cures? Tracing Assetization and Value in Biomedical Innovation. *Assetization: Turning Things into Assets in Technoscientific Capitalism*, pp. 89–116. New York: MIT Press.

Rutherford, S. and Rutherford, P. 2013. Geography and Biopolitics. *Geography Compass*, 7(6), pp. 423–434.

Sample, I. 2021. JCVI Failed to Back Youth Covid Jabs Despite Favourable Modelling. *The Guardian*, November 3. https://www.theguardian.com/politics/2021/nov/02/jcvi-failed-back-youth-covid-jabs-despite-favourable-models.

Sankar P. 2010. Forensic DNA phenotyping: reinforcing race in law enforcement. *What's the Use of Race? Modern Governance and the Biology of Difference* I Whitmarsh, DS Jones 49–61 Cambridge, MA: MIT Press

Sankar, P. and Kahn, J. 2005. BiDil: Race Medicine or Race Marketing? *Health Affairs, Web Exclusive*, W5- 455-W5-463; Kahn, J. 2005. BiDil: False Promises. *GeneWatch*, 18(6): e-version; Kahn, J. 2008. Exploiting Race in Drug Development: BiDil's Interim Model of Pharmacogenomics. *Social Studies of Science*, 38(5), pp. 737–758.

Santos, B. D. S. 2015. Towards a Socio-Legal Theory of Indignation. In *Law's Ethical, Global and Theoretical Contexts*, pp. 115–142. Cambridge: Cambridge University Press.

Sapolsky, Robert. 2004. The Frontal Cortex and the Criminal Justice System. *The Royal Society*, 359(145), p. 1787.

Schafft, Gretchen E. 2002. Scientific Racism in Service of the Reich: German Anthropologists in the Nazi Era. *Annihilating Difference: The Anthropology of Genocide*, pp. 117–136. Berkeley: University of California Press.

Schiff, A. R. 1994. Solomonic Decisions in Egg Donation: Unscrambling the Conundrum of Legal Maternity. *Iowa Law Review*, 80, p. 265.

Schwartz, Richard D. 1982. On the Prospects of Using Sociobiology in Shaping the Law: A Cautionary Note. *Journal of Social and Biological Structures*, 5(4), pp. 325–333.

Schwartz-Marín, Ernesto and Méndez, Alberto Arellano. 2012. The Law of Genomic Sovereignty and the Protection of Mexican Genetic Patrimony. *Medicine and Law*, 31, p. 283.

Scott, J. C. 1998. *Seeing Like a State: How Certain Schemes to Improve the Human Condition Have Failed*. New Haven, CT: Yale University Press.

Sell, S. 2003. *Private Power, Public Law: The Globalization of Intellectual Property Rights*. Cambridge: Cambridge University Press.

Selznick, P. 1961. Sociology and Natural Law. *Natural Law Forum*, 6, pp. 84–108.

Settumba, S., Butler, T., Schofield, P., Chambers, G. M., and Shanahan, M. 2021. Societal Preferences for the Treatment of Impulsive-Violent Offenders: A Discrete Choice Experiment. *BMJ Open*, 11. https://doi.org/10.1136/bmjopen-2019-033935.

Shapiro, S. J. 2011. *Legality*. Cambridge: Harvard University Press.

Shen, F. X. 2016. The Overlooked History of Neurolaw. *Fordham Law Review*, 85, pp. 667–695.

Sherman, B. 2008. Taxonomic Property. *The Cambridge Law Journal*, 67(3), pp. 560–584.

Shildrick, M. 2015. Chimerism and Immunitas: The Emergence of a Posthumanist Biophilosophy. In *Resisting Biopolitics*, pp. 107–120. London: Routledge.

Shildrick, M., 2022. Maternal–Fetal Microchimerism and Genetic Origins: Some Socio-legal Implications. *Science, Technology, & Human Values, 47*(6), pp.1231–1252.

Silbey, S. S. 1992. Making a Place for Cultural Analyses of Law: [Commentary]. *Law & Social Inquiry*, 17(1), pp. 39–48.

Silbey, S. S. 2005a. Everyday Life and the Constitution of Legality. In Jacobs, M. and Hanrahan, N. (eds.), *The Blackwell Companion to the Sociology of Culture*, pp. 332–345. Oxford: Blackwell.

Silbey, S. S. 2005b. After Legal Consciousness. *Annual Review of Law and Social Science*, 1, pp. 323–368.

Silbey, S. S. 2011. J. Locke, op. cit.: Invocations of Law on Snowy Streets. *Journal of Comparative Law*, 5(2), pp. 66–91.

Simpson, B. 2013. Managing Potential in Assisted Reproductive Technologies: Reflections on Gifts, Kinship, and the Process of Vernacularization. *Current Anthropology*, 54(S7), pp. S87–S96.

Sirgiovanni, Elisabetta and Garasic, Mirko Daniel. 2020. Commentary: The Moral Bioenhancement of Psychopaths. *Frontiers in Psychology*, 10, p. 2880.

Skloot, R. 2010. *The Immortal Life of Henrietta Lacks*. New York: Crown Publishers.

Slabbert, M. Nothling and Pepper, Michael S. 2010. A Room of Our Own: Legal Lacunae Regarding Genomic Sovereignty in South Africa. *THRHR*, 73, p. 432.

Slaby, J. and Choudhury, S. 2018. Proposal for a Critical Neuroscience. In Meloni, M., Cromby, J., Fitzgerald, D., and Lloyd, S. (eds.), *The Palgrave Handbook of Biology and Society*, pp. 341–370. London: Palgrave Macmillan.

Slaby, J. and Gallagher, S. 2015. Critical Neuroscience and Socially Extended Minds. *Theory, Culture and Society*, 32(1), pp. 33–59.

Sparrow, R. and Mills, C. 2021. Genome Editing: From Bioethics to Biopolitics. *BioSocieties*, 17(3), pp. 485–505.

Spenser, K., Bull, R., Betts, L., et al. 2019. Executive Functioning as a Predictive Measure of Offending Behaviour. *Journal of Criminal Psychology*, 9(1), pp. 10–22.

Stallins, J. Anthony, Law, Derek M., Strosberg, Sophia A., and Rossi, J. J. 2018. Geography and Postgenomics: How Space and Place are the New DNA. *GeoJournal*, 83(1), pp. 153–168.

Stern, Claudio D. 2019. The 'Omics Revolution: How an Obsession with Compiling Lists Is Threatening the Ancient Art of Experimental Design. *BioEssays*, 41(12), n/a–n/a. Web.

Stevens, H. 2016. *Biotechnology and Society: An Introduction*. Chicago: University of Chicago Press.

Strathern, M. 1996. Enabling identity? Biology, choice and the new reproductive technologies. In Hall, S. and Du Gay, P. *Questions of cultural identity*, PP. 37–45, London: Sage

Strathern, M. 1998. Gender: Division or Comparison? *The Sociological Review* 45(1_suppl), pp. 42–63.

Strathern, M. 1999. *Property, Substance, and Effect: Anthropological Essays on Persons and Things*. London: Athlone Press.

Strathern M. 2001. The Patent and the Malanggan. *Theory Cult. Soc.* 18(4):1–26

Strathern, M. 2005. *Kinship, Law and the Unexpected: Relatives Are Always a Surprise*. Cambridge: Cambridge University Press.

Sun, S. 2017. Socio-economics of personalized medicine in Asia, Routledge.

Swiffen, A. 2010. *Law, Ethics and the Biopolitical*. London: Routledge.

TallBear, K. 2013. Native American DNA: Tribal Belonging and the False Promise of Genetic Science. Minnesota: University of Minnesota Press.

Taussig, K.S., Hoeyer, K. and Helmreich, S., 2013. The anthropology of potentiality in biomedicine: an introduction to supplement 7. *Current Anthropology*, 54(S7), pp.S3–S14.

Taylor, J. S. 1995. Neurolaw: Towards a New Medical Jurisprudence. *Brain Injury*, 9(7), pp. 745–751.

Taylor, J. S. 1997. *Neurolaw: Brain and Spinal Cord Injuries*. Houston: Thompson Legal Publishing.

Teubner, G. 1993. *Law as an Autopoietic System*. Oxford and Cambridge: Blackwell Publishers.

Teubner, G. 2012. *Constitutional Fragments: Societal Constitutionalism and Globalization*. Oxford: Oxford University Press.

Thacker, E. 2006. *The global genome: Biotechnology, politics, and culture*. MIT press.

Thompson, C. 2009. Kinship in an Infertility Clinic. *Feminist Anthropology: A Reader*, p. 271. Malden: Wiley-Blackwell.

Timmermans, S. and Buchbinder, M. 2010. Patients-in-waiting: living between sickness and health in the genomics era. *Journal of health and social behavior*, 51(4), pp. 408–423.

Trimmings, K. and Beaumont, P. 2011. International Surrogacy Arrangements: An Urgent Need for Legal Regulation at International Level. *Journal of Private International Law*, 7, pp. 627–647.

Tsing, A.L. 2005. *Friction: An Ethnography of Global Connection*. Princeton University Press.

Tsing, A. L. 2015. *The Mushroom at the End of the World*. Princeton: Princeton University Press.

Valdez, N., 2018. The redistribution of reproductive responsibility: On the epigenetics of "environment" in prenatal interventions. *Medical Anthropology Quarterly*, 32(3), pp. 425–442.

Van Beers, B. 2016. From Winged Lions to Frozen Embryos, Neomorts and Human-Animal Cybrids: The Functions of Law in the Symbolic Mediation of Biomedical Hybrids. In *Symbolic Legislation Theory and Developments in Biolaw*, pp. 177–199. Cham: Springer.

Van Beers, B. 2017. The Changing Nature of Law's Natural Person: The Impact of Emerging Technologies on the Legal Concept of the Person. *German Law Journal, 18*(3), 559–594.

Van Beers, B. C. 2018. The Best Me I Can Possibly Be: Legal Subjectivity, Self-Authorship and Wrongful Life Actions in an Age of 'Genomic Torts'. In *Personalised Medicine, Individual Choice and the Common Good*, pp. 200–225. Cambridge: Cambridge University Press.

Van Beers, B.C. 2020. Rewriting the Human Genome, Rewriting Human Rights Law? Human Rights, Human Dignity, and Human Germline Modification in the CRISPR Era. *Journal of Law and the Biosciences*, 7(1), pp. 1–36.

van Beers, B., Corrias, L., and Werner, W. G., eds. 2014. *Humanity Across International Law and Biolaw*. Cambridge: Cambridge University Press.

Van Beers B, van Beers BC, Sterckx S, Dickenson D, (eds) 2018. *Personalised medicine, individual choice and the common good*. Cambridge University Press.

Van de Wiel, L. 2015. Frozen in Anticipation: Eggs for Later. *Women's Studies International Forum*, 53, pp. 119–128.

Van de Wiel, L. 2020. *Freezing Fertility: Oocyte Cryopreservation and the Gender Politics of Aging*. New York: New York University Press.

Van Klink, Bart, Van Beers, Britta, and Poort, Lonneke, eds. 2016. *Symbolic Legislation Theory and Developments in Biolaw*. Springer: Springer International Publishing.

Van Wichelen, S. 2016. Postgenomics and Biolegitimacy: Legitimation Work in Transnational Surrogacy. *Australian Feminist Studies*, 31(88), pp. 172–186.

Van Wichelen, S. 2019a. *Legitimating Life: Adoption in the Age of Globalization and Biotechnology*. New Brunswick: Rutgers University Press.

Van Wichelen, S. 2019b. Law's Relations in the Regulation of Cross-Border Surrogacy. In Mackie, Vera, Marks, Nicola, and Ferber, Sarah (eds.), *The Reproductive Industry: Intimate Experiences and Global Processes.* Lanham: Lexington Books.

Van Wichelen, S. 2019c. Moving Children Through Private International Law: Institutions and the Enactment of Ethics. *Law & Society Review*, 53(3), pp. 671–705.

van Wichelen, S. 2022. Identity in postgenomic times: Epigenetic knowledge and the pursuit of biological origins. *Science, Technology, & Human Values*, 47(6), pp.1131–1156.

Van Wichelen, S. 2023. After Biosovereignty: The Material Transfer Agreement as Technology of Relations. *Social Studies of Science*, 53(4), pp. 599–621.

Van Wichelen, S. and Keaney, J. 2022. The Reproductive Bodies of Postgenomics. *Science, Technology, & Human Values*, 47(6), pp. 1111–1130.

Vatter, M. 2022. Planetary Health and the Biopolitics of Home. In *The Viral Politics of Covid-19: Nature, Home, and Planetary Health*, pp. 221–246. Cham: Palgrave Macmillan.

Vatter, Miguel and De Leeuw, Marc. 2023. Human Rights, Legal Personhood and the Impersonality of Embodied Life. *Law, Culture and the Humanities*, 19(1), pp. 106–125.

Vidal, Fernando. 2009. Brainhood, Anthropological Figure of Modernity. *History of the Human Sciences*, 22(1), p. 7.

Vincent, Nicole A. 2013. *Neuroscience and Legal Responsibility*. Oxford: Oxford University Press.

Virruel-Fuentes, E. A. (2007). Beyond acculturation: Immigration, discrimination, and health research among Mexicans in the United States. *Social Science & Medicine*, 65, 1524–1535.

Vismann, C. 2013. Cultural Techniques and Sovereignty. *Theory, Culture & Society*, 30(6), 83–93.

Vora, K. 2015. Re-imagining Reproduction: Unsettling Metaphors in the History of Imperial Science and Commercial Surrogacy in India. *Somatechnics*, 5(1), pp. 88–103.

Waldby, C. 2015. 'Banking Time': Egg Freezing and the Negotiation of Future Fertility. *Culture, Health & Sexuality*, 17(4), pp. 470–482.

Waldby, C. and Cooper, M. 2008. The Biopolitics of Reproduction: Post-Fordist Biotechnology and Women's Clinical Labour. *Australian Feminist Studies*, 23(55), pp. 57–73.

Waldby, C. and Mitchell, R. 2006. *Tissue economies: Blood, organs, and cell lines in late capitalism*. Duke University Press.

Wardlow v. Director, CIVIL ACTION NO. 4:04-CV-408, Eastern District of Texas (E.D. Tex.). 2017, August 21.

Wardlow v. Texas, 19–8835, (19A1065), Supreme Court of United States. 2020, July 8.

Warin, M. Kowal, E. and Meloni, M. 2020. Indigenous knowledge in a postgenomic landscape: The politics of epigenetic hope and reparation in Australia. *Science, Technology, & Human Values*, 45(1), pp.87–111.

Watt, H. M. 2011. Private International Law Beyond the Schism. *Transnational Legal Theory*, 2(3), pp. 347–428.

Watt, H. M. 2016. Conflicts of Laws Unbounded: The Case for a Legal-Pluralist Revival. *Transnational Legal Theory*, 7(3), pp. 313–353.

Whitmarsh, Ian. 2009. Hyperdiagnostics: Postcolonial Utopics of Race-Based Biomedicine. *Medical Anthropology*, 28(3), pp. 285–315.

Whitmarsh, Ian and Jones, David S., eds. 2010. *What's the Use of Race? Modern Governance and the Biology of Difference*. New York: MIT Press.

Whittaker, A. 2018. *International Surrogacy as Disruptive Industry in Southeast Asia*. New Brunswick: Rutgers University Press.

Wilson, E. A. 2015. *Gut feminism*. Duke University Press.

Winickoff, D. E. 2015. 2 Biology Denatured: The Public-Lives of Lively Things. In Hilgartner, S., Miller, C. A., and Hagendijk, R. (eds.), *Science and Democracy: Making Knowledge and Making Power in the Biosciences and Beyond*, p. 15. London: Routledge.

Winickoff, D. E. and Obasogie, O. 2008. Race-Specific Drugs: Regulatory Trends and Public Policy. *Trends in Pharmacological Sciences*, 29(6), pp. 277–279.

Yngvesson, B. 2010. *Belonging in an Adopted World: Race, Identity, and Transnational Adoption*. Chicago: University of Chicago Press.

Young, Iris Marion. 1990. *Justice and the Politics of Difference*. Princeton: Princeton University Press.

Zafran, R. 2007. Dying to be a Father: Legal Paternity in Cases of Posthumous Conception. *Houston Journal of Health Law & Policy*, 8, p. 47.

Court References

In re Baby M, 537 A.2d 1227, 109 N.J. 396 (N.J. 1988)

re Marriage of Buzzanca, 61 Cal. App. 4th 1410, 72 Cal. Rptr. 2d 280, 1998 Cal. App. LEXIS 180, 98 Cal. Daily Op. Service 1782, 77 A.L.R.5th 775, 98 Daily Journal DAR 2436 (Cal. App. 4th Dist. Mar. 10, 1998).

R v. Human Fertilisation and Embryology Authority, ex parte Blood, see also Dyer 2013.

R (Mr and Mrs M) v Human Fertilisation and Embryology Authority, [2016] EWCA Civ 611.

Case of Mennesson v. France, Application n. 65192/11, Judgment 26 June 2014, and Case of Labassee v. France, Application n. 65941/11, Judgment 26 June 2014.

Index

S. van Wichelen and M. de Leeuw, *Biolegality*, Biolegalities,
https://doi.org/10.1007/978-981-99-8749-8

FSC
www.fsc.org
MIX
Papier | Fördert gute Waldnutzung
FSC® C083411

Kolibri 360 GmbH
Ferdinand-Jühlke-Straße 7
99095 Erfurt, Deutschland
produktsicherheit@kolibri360.de